**Praise for *The Trend Management Toolkit***

"This book is truly inspirational. An excellent toolkit for anybody who needs to think about what the future will hold and prepare their business for it before it happens. It will help all its readers both to see and understand 'the big picture' and how we will evolve as people, consumers, and societies in a not so distant future."

John Carstensen, Head of Profession for Climate and Environment, UK Department for International Development

"I've read this book with the utmost interest and found it not only a professionally useful but a personally captivating read as well. It offers plenty of food for thought and – what is more important – a methodology and a set of tools that really help to organize and orient the reflection process, be it individual or collective. *The Trend Management Toolkit* offers a rare combination of sharp insights, sound methodology, and effective tools. It will be of great help to companies when it comes to navigating the complexity of the future."

Fernando Gutiérrez, Director and Advisor to the Chairman and CEO, BBVA

"To make our world a better and greener place, we need to work together for the common good. This book presents the business case for a balanced people, planet, purpose, and profit outlook – a sustainable business model that is close to my own heart. It is essential reading for anyone wishing to map out and create positive future scenarios."

Søren Hermansen, CEO, Samsø Energy Academy; *TIME* magazine Hero of the Environment 2008

"As we are living in a world where the amount of data is increasing dramatically, technology is developing at a speed never seen before, and the stress we are putting on environment and resources is coming to a tipping point – it is hard not to get confused. But with *The Trend Management Toolkit* we get sound help to navigate in this complex world."

Anders Eldrup, Former CEO, DONG Energy; Permanent Secretary, Danish Ministry of Finance

"Anne Lise Kjaer provides a valuable toolkit for conscious capitalists looking to shape the 21st century. Unapologetically Nordic in her approach, she shows how a strong focus on inclusion, empowerment, and social consciousness can spur innovation and give businesses a competitive edge. This book is a must-read for current and future business leaders."

Henrik Fogh Rasmussen, Founder, Rasmussen Public Affairs

"This inspiring and generous book cleverly explains how society has developed over the last 20 years and pulls all the complexity into an accessible tool that ensures strategic thinking. Dig in and choose what future you want to create – the business case is already argued thoroughly and the warmth and wit make it a captivating read. As both a toolbox and a summary of 20 years plus of professional learning, I can see it becoming a key source for university courses, not just within business but also design and sociology. It will also be the Christmas present for my entire network."

Esther Davidsen, Brussels lobbyist; head of Zealand Denmark EU Office

"Trends on their own tell a small part of the story. Using techniques from *The Trend Management Toolkit*, business leaders can apply a systemic approach to understanding how to visualize the future."

R "Ray" Wang, Founder, Constellation Research Inc.;
author of *Disrupting Digital Business*

"Our success relies on adapting to the future faster than our competitors. This book's toolkit is an invaluable resource to think deeply about that future."

Sir Ian Cheshire, Group CEO, Kingfisher plc

"Just as Alvin Toffler's *Third Wave* helped frame the coming decades for me in 1980, Anne Lise Kjaer has put words to the uncertainties and opportunities facing our global village in coming years. The need to think in terms of circular economies and understand the critical relevance of social inclusion has never been greater. She equips the reader with both the mindset and the tools to help adapt, and even thrive, in the face of radical change."

Gary Baker, Executive Director, Climate Change &
Sustainability Services, Ernst & Young AB

"The future, and how to make sense of it – these are the themes of Anne Lise Kjaer's new book. A guidebook to long-term planning in a fast-morphing world, *The Trend Management Toolkit* will particularly interest businesses looking for new ways of responding to changes in consumer behaviour. But its potted history of the art of prediction will fascinate all – as should Kjaer's argument that vapid consumerism has heightened, rather than destroyed, a desire for social ties and collaboration."

Patrick Kingsley, *Guardian* foreign correspondent, author of *How to be Danish*; best young journalist at the 2014 British Press Awards

"Trends – every company depends on understanding them yet few do. In Anne Lise Kjaer's new book, *The Trend Management Toolkit*, she takes a brave step into the future by introducing a new way of predicting and managing trends. She has cracked it!"

Martin Lindstrom, *New York Times* bestselling author of *Brandwashed* and *Buyology*

"The opportunities presented by the ongoing digital and social revolution make a culture and architecture of continuous innovation an even bigger imperative for progressive organisations. The ability to paint a holistic picture of the future is an important component of this innovation architecture. *The Trend Management Toolkit* provides a valuable method to map the future and thereby contribute directly to the innovation process."

Neetan Chopra, Senior Vice President, IT Strategy, Emirates Group

"I was captivated by the direct, personal and inspiring tone of this toolkit and storytelling book about the future. I read it in one day – simply couldn't put it down – because I wanted to know more. A new societal learning map and master class is born, complete with history, data and tools. This is a solid overview, with clear arguments and insights, which explains why we need a new and multidimensional set of tools to navigate in the 21st century.

Lars Engman, former design director, IKEA

"The future is something to be embraced not feared, but it has never been harder to discern. Political, economic, social, environmental, and technology trends bounce around the world, colliding and merging at a fearsome rate; Anne Lise's book is a great practical guide to understanding these trends and building a strategy to respond to them."

Mike Barry, Director Plan A, Marks & Spencer

"In an era of constant change, brands are finding that it is vital to be able to identify and embrace new and game-changing trends. *The Trend Management Toolkit* offers incisive insights and practical steps to help marketers better understand the future needs and wants of their customers."

Muireann Bolger, Features Editor, *The Marketer*

# The Trend Management Toolkit

## A Practical Guide to the Future

Anne Lise Kjaer

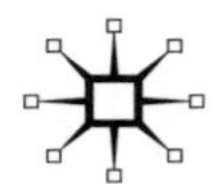



First published 2014 by
PALGRAVE MACMILLAN

Palgrave Macmillan in the UK is an imprint of Macmillan Publishers Limited, registered in England, company number 785998, of Houndmills, Basingstoke, Hampshire RG21 6XS.

Palgrave Macmillan in the US is a division of St Martin's Press LLC, 175 Fifth Avenue, New York, NY 10010.

Palgrave Macmillan is the global academic imprint of the above companies and has companies and representatives throughout the world.

Palgrave® and Macmillan® are registered trademarks in the United States, the United Kingdom, Europe and other countries.

ISBN: 978–1–137–37008–2

This book is printed on paper suitable for recycling and made from fully managed and sustained forest sources. Logging, pulping and manufacturing processes are expected to conform to the environmental regulations of the country of origin.

A catalogue record for this book is available from the British Library.

A catalog record for this book is available from the Library of Congress.

Typeset by Aardvark Editorial Limited, Metfield, Suffolk.

*This book is dedicated to my son*
*Vicente Macia-Kjaer who challenges*
*and inspires me by bringing a true*
*Millennial dimension to every discourse*
*about the future.*

# Contents

# List of figures and tables

## Figures

## Tables

# Foreword

We are faced with a particularly tough set of challenges in today's information society. With a vast and growing data deluge, keeping up with analysis, pattern spotting, and extracting useful information has never been more crucial; hence, we need a system that talks back to us and enables informed decision making. Over the past 20 years, I have developed tools and approaches to help organizations discuss and determine the best way ahead. The Trend Atlas integrates the management of data and societal insights to provide a sense-making platform for anticipating future challenges and opportunities.

My roots in Denmark, where democratic thinking and a strong design culture are ingrained, have profoundly influenced how I navigate the world and map "what comes next." Visiting Copenhagen, the capital of one of world's happiest nations, always inspires me. Here, it appears to me, the future is already happening. Urbanization in Denmark stands at 87%, yet over 40% of the capital's commuters travel by bicycle and the country has robust plans to run solely on renewable energy by 2050. Strongly committed to e-government, it has 100% broadband penetration. Employment rates are high, four out of five women work, and the affordable kindergarten system, attended by 90% of Danish children, is a highly visible investment in nurturing tomorrow's smart and inclusive society.

As you will realize, I am an unapologetic believer in the values I grew up with. Recognizing that there is not just one but many possible futures, we can't ignore that our choices today profoundly influence tomorrow. When we consider the full range of societal values, we enable groundbreaking scenario thinking. It is in this spirit that I set out my road map for how trend management delivers value to business and society as a whole.

# Acknowledgments

This book simply wouldn't have become a reality without all the inspiring people who helped me develop it.

First of all, a huge thank you to Libby Norman, my editor since 2003, who carefully edited each chapter to perfection and summarized sections into digestible "future sound bites." Sociologist Esther Giner Macia meticulously scanned and condensed more than a decade of Kjaer Global's work, and helped research the core concepts presented in this book, while my husband, architect Harald Brekke, lent his analytical and critical judgment, tirelessly reading the manuscript again and again. I am indebted to Louise Loecke Foverskov, design and trend strategist at Kjaer Global since 2006, who gave invaluable input over the years, not least our work to evolve the toolkit shown here. Environment strategist Ela Rose diligently assisted in researching the trends and refining the case studies. All of you were instrumental in bringing this book to fruition. Also, a special thanks to my publishers Palgrave Macmillan and the Aardvark Editorial team for their extraordinary work during the development, editing, and production phases.

Last, but not least, I could not have amassed the experience or knowledge to write this book without the support from Kjaer Global's clients, who have given us the great privilege to share their journeys into the future. While there are too many great brand names to list here, they know who they are. All of our clients – through their visionary leadership – demonstrate time and again how strategies that focus on improving lives and communities will also deliver on the bottom line.

**Anne Lise Kjaer**
Futurist and Copenhagen Goodwill Ambassador

Learn more on **www.kjaer-global.com**

# Chapter 1

# From facts to feelings

*In the past, when I mentioned "feelings" to companies the immediate response was: "We are not interested in feelings about the future, we want facts."*

Nobel Prize-winning psychologist Daniel Kahneman, who won his prize in economics, noted in an interview for UK newspaper *The Guardian* in 2012:[1]

> Many people now say they knew a financial crisis was coming, but they didn't really. After a crisis we tell ourselves we understand why it happened and maintain the illusion that the world is understandable. In fact, we should accept the world is incomprehensible much of the time.

While I agree with part of Kahneman's statement, I also believe that very often the writing is on the wall before seismic change occurs – we simply need a method of tuning into the vibrations of our universe. This means that, alongside the gathering of data and insights to give us rational knowledge, we must develop our intuition to enable us to predict and develop strategies for change.

## Introducing trend management as a concept

To make sense of the future, even the seemingly sudden shifts, we must have a system in place that sifts and correlates trends. The Trend Management Toolkit is such a system, acting as a platform for integrated thinking that allows us to anticipate developments and make more informed choices in the present about the future. At its simplest, the process of managing trends involves observing specific changes or advances, as well as considering the general direction in which society is moving. But to do this consistently and with confidence, it is essential to have a trend management system that deals with the complexity of diverse information, fosters alternative viewpoints, and generates fresh thoughts. This offers a structure that makes trend forecasting come alive, enabling us to discover and analyze trends that tell us something about the future as well as inspiring timely ideas and solutions.

The system I describe in this book is a powerful strategic method developed for making sense of the many multifaceted, sometimes conflicting, drivers influencing today's reality and tomorrow's world. Trend mapping and scenario building are typically part of the trend management process, in which we build narratives around nascent trends to consider likely outcomes for organizations by projecting the impact for business development, product design, and service concepts. We use a wide range of data for broad insight, incorporating experts' opinions as well as case studies and analysis to build a 360-degree outlook. The outcome is a framework that can be applied to everything from new business models, innovation and design strategies to brand building and marketing.

## Decoding society and human behavior

Future forecasting is a relatively new and developing field. However, in recent decades, it has become an increasingly widely accepted decision-making tool for assessing societal influences, economic drivers, and – ultimately – boosting sales to increase revenues and reputation. Over the past 20 years, my team and I have built tools and processes to help companies and organizations navigate the future, often refining and developing these tools in response to the specific challenges faced by the organizations we've worked with, as well as taking on board new theories and approaches as we've added fresh research perspectives to our multidisciplinary practice. Today, this methodology allows us to combine a wide spectrum of trend snapshots to create viable and inspirational scenarios.

One of our vital tools is the Trend Atlas, a visual sense-making platform that integrates management of a wide variety of data and information to provide insights for determining what lies ahead. The Trend Atlas is a structured compilation of macro trends that acts as a compass, enabling us to decode the socioeconomic and cultural contexts of society to decipher patterns that provide a framework for projection, planning, and ideation. Trend management combines a wide spectrum of drivers and insights to create powerful future sound bites. These are essentially the building blocks for creating sustainable and credible future narratives that make it possible for companies to explore potential developments, both short to medium term. The narratives are underpinned by a variety of research findings and insights – not just numbers – and enable us to contextualize lifestyle situations that consider people's future preferences, choices, and actions. In effect, we are teleported into the future and encouraged to ask the big "what if?" questions. The narratives are more compelling than a simple forecast and allow organizations to visualize future situations in a believable, multilayered way. As such, they become powerful tools for imagining the future and creating sound strategies, as well as managing risk.

The sociology of people is an essential component in understanding the future, but we also need to factor in the sociology of things – technology in particular, as technological development plays a key role in the interpretation of how global economies and cultures connect. This is another key reason why multidimensional trend management is fundamental for imagining the future. Its much broader set of research tools invite us to detach ourselves from our current, local context, consider the whole picture and thereby view our organization holistically to understand how we come across to the rest of the world – a process we call "looking from the outside in." In order to assist our clients develop their critical thinking about the future, we also consider the evolution of societies, businesses, lifestyle patterns, and the environment. We find that when we observe the past and present, it's possible to gain deeper insights into how the future might unfold. In a nutshell, trend management assists with the process of mapping out current trends and influences for businesses within a society-wide context.

## The evolution of future studies

The business of trend forecasting and the need for it is nothing new; indeed, there have always been thinkers and seers imagining the future (see Table 1.1). However, it is only in the last half-century that it has become a discernible business with its own distinct methodologies and

TABLE 1.1 **Some key influencers through time**

| Year | Name | Expertise and future-focused studies |
|---|---|---|
| 1452–1519 | Leonardo da Vinci<br>Italian | Artist, engineer, scientist and inventor, architect, musician, mathematician, anatomist, geologist, cartographer, botanist, and writer. Surviving notebooks reveal the most eclectic of predictive minds. Key publications: *Codex Arundel* (1480), *Codex Leicester* (1508), *Codex Atlanticus* (1519) |
| 1828–1905 | Jules Verne<br>French | Novelist, poet, and playwright. Inspired many future scientists with his body of science fiction collected within *Voyages Extraordinaires* (1863–1905). Key publications: *Journey to the Centre of the Earth* (1864), *From the Earth to the Moon* (1865), *Twenty Thousand Leagues Under the Sea* (1870), *Around the World in Eighty Days* (1873) |
| 1856–1943 | Nikola Tesla<br>Serbian-American | Futurist inventor, electrical mechanical engineer, and physicist. Devised the modern alternating current electricity supply system, with 300 inventions patented worldwide and amazing 21st-century predictions |
| 1866–1946 | H.G. Wells<br>English | Futurist and science fiction author, with wide-ranging interests in science and social policy. Considered among the founders of future studies. Key publications: *The Time Machine* (1895), *The War of the Worlds* (1898), *Anticipations* (1901), *A Modern Utopia* (1905), *The Shape of Things to Come* (1933), *World Brain* (1936–38) |
| 1895–1983 | Buckminster Fuller<br>American | Futurist, architect, systems theorist, author, designer, and inventor. Key publications: *4D Timelock* (1928), *Operating Manual for Spaceship Earth* (1968), *Critical Path* (1981) |
| 1896–1960 | Gaston Berger<br>French | Futurist, industrialist, philosopher, and modernizer of French university system. Contributed noted analysis of Edmund Husserl's work. Key publication: *Recherches sur les conditions de la connaissance* (1941) |
| 1900–86 | Walter Greiling<br>German | Futurist and chemist. Researched agricultural microbiology and predicted that systematic international efforts to mitigate climate change would begin in 1990. Key publication: *Wie werden wir leben? Ein Buch von den Aufgaben unserer Zeit* (1954) |
| 1903–87 | Bertrand de Jouvenel<br>French | Futurist, political economist, philosopher, and author. Pioneer of future studies. Key publications: *On Power: The Natural History of Its Growth* (1945), *The Ethics of Redistribution* (1951) |
| 1906–92 | Grace Hopper<br>American | Mathematician, computer scientist, and US Navy rear admiral. One of the first programmers of the Harvard Mark I computer and developed the first compiler for computer programming language |

| Year | Name | Expertise and future-focused studies |
|---|---|---|
| 1907–85 | Fred Polak<br>Dutch | Futurist, philosopher, and sociologist. Theorized the central role of imagined alternative futures. Key publication: *The Image of the Future* (1973) |
| 1911–80 | Marshall McLuhan<br>Canadian | Futurist, media theorist, philosopher, and author. Coined the phrases "the medium is the message" and "the global village" and predicted the World Wide Web in the 1960s. Key publications: *The Gutenberg Galaxy* (1962), *Understanding Media* (1964), *Medium is the Massage: An Inventory of Effects* (1967), *The Global Village: Transformations in World Life and Media in the 21st Century* (1989) |
| 1915–69 | M.G. Gordon<br>American | Futurist, businessman, inventor, and social theorist. Advocate for privacy rights and envisioned expanded telephone network as the ideal social network |
| 1916– | Jacque Fresco<br>American | Structural engineer, architectural designer, author, and educator. Writes and lectures on sustainable cities, energy efficiency, natural resource management, cybernetic technology, advanced automation, science in society |
| 1917–2008 | Arthur C. Clarke<br>British | Futurist, inventor, TV presenter, and science fiction writer. A polymath and popularizer of science, with many notable works about the future. The book and film (with Stanley Kubrick) *2001: A Space Odyssey*, based partly on two earlier short stories by Clarke, is one of the most influential films of all time |
| 1918–88 | Richard Feynman<br>American | Theoretical physicist. Introduced the concept of "nanotechnology" and received the Nobel Prize in Physics in 1965 |
| 1919– | James Lovelock<br>British | Futurist, independent scientist, and environmentalist. Proposed the Gaia hypothesis and the concept "sustainable retreat." Key publication: *Gaia: A New Look at Life on Earth* (1979) |
| 1919–2009 | Russell L. Ackoff<br>American | Organizational theorist, consultant, professor. Pioneer in the field of operations research, systems thinking and management science, inspiring many future developments in areas such as decision science |
| 1920–92 | Isaac Asimov<br>American | Author and professor of biochemistry. Known for his works of science fiction and science. Key publications: *The Foundation Series* (1949–93), *I, Robot* (1950), *The Intelligent Man's Guide to Science* (1960) |
| 1921–2006 | Stanisław Lem<br>Polish | Writer of science fiction, philosophy, and satire. His books explore philosophical themes, including the nature of intelligence, communication, human limitations, technology, and the universe. Key publications: *Solaris* (1961), *The Cyberiad* (1965) |

| Year | Name | Expertise and future-focused studies |
|---|---|---|
| 1922–83 | Herman Kahn American | Futurist, physicist, mathematician, and thinker. A military strategist and systems theorist who analyzed the likely consequences of nuclear war. Key publications: *On Thermonuclear War* (1960), *The Coming Boom: Economic, Political and Social* (1983) |
| 1922–97 | Pierre Wack French | Oil executive. Developed the use of scenario planning in the private sector in the 1970s. His articles are among the first to bring Herman Kahn's theories into business strategy |
| 1923– | Freeman Dyson British-American | Theoretical physicist and mathematician. Pioneering work in quantum electrodynamics, solid-state physics, astronomy, and nuclear engineering |
| 1924–2010 | Benoît B. Mandelbrott French-American (born Poland) | Mathematician. The founder of fractal geometry to understand roughness in nature and complex data sets. An ambassador for "the unity of knowing and feeling." Key publications: *Les objets fractals* ([1975] 1995), *The (Mis)Behaviour of Markets: A Fractal View of Risk, Ruin and Reward* (2004) |
| 1925–2013 | Douglas Engelbart American | Engineer, inventor, computer and Internet pioneer. Focused on the area of human–computer interaction in order to improve computer interfaces long before PCs were even envisaged |
| 1928– | Alvin Toffler American | Futurist, author, and inventor. Exploratory work mapping the digital and communication revolutions and technological singularity. Key publications: *Future Shock* (1970), *The Third Wave* (1980) |
| 1929– | John Naisbitt American | Futurist and author. Expanded interest in trends and coined the phrase "radical center." Key publication: *Megatrends* (1982) |
| 1934–96 | Carl Sagan American | Astronomer, author, and science popularizer. Key publications: *Cosmos* (1980), *Contact* (1997) |
| 1937– | Joël de Rosnay French (born Mauritius) | Futurist, science writer, and molecular biologist. Pioneered the role of the Internet in the emergence of a global brain |
| 1945– | Jeremy Rifkin American | Political advisor, economic and social theorist. President of Foundation on Economic Trends with key focus on the impact of scientific and technological changes. Key publications: *The End of Work* (1995), *The Age of Access* (2000), *The Empathic Civilization* (2010) |
| 1945– | Jerome C. Glenn British | Futurist, co-founder of The Millennium Project. Created the Futures wheel technique and is a Singularity University advisor. Key publication: *Futures Research Methodology Version 3.0* (2009) |

| Year | Name | Expertise and future-focused studies |
|---|---|---|
| 1947– | Michio Kaku American | Futurist, theoretical physicist, and communicator. University professor and popularizer of science topics. Key publications: *Physics of the Impossible* (2008), *Physics of the Future* (2011) |
| 1947– | Hugo de Garis Australian | Professor and researcher of artificial intelligence. Noted for work in field of "intelligent machines" and has attracted controversy for belief that major conflict will be a by-product of their evolution. Key publication: *Artificial Brains: An Evolved Neural Net Module Approach* (2009) |
| 1948– | Ray Kurzweil American | Futurist, inventor, computer scientist, and author. Involved in development of several technologies, including CCD flatbed scanner, text-to-speech synthesizer, and first commercially marketed large-vocabulary speech recognition. Key publications: *The Singularity is Near* (2005), *How to Create a Mind: The Secret of Human Thought Revealed* (2012) |
| 1950– | Lidewij Edelkoort Dutch | Trend forecaster, designer, publisher. Influential developer of brand and product identities via lifestyle analysis, trend mood books, and audiovisuals. Key publications: *View on Colour* magazine |
| 1952– | Kevin Kelly American | Futurist, writer, editor, publisher, and photographer. Co-founded *Wired* magazine in 1993. Key publications: *Signal: A Whole Earth Catalog* (1988), *Out of Control: The New Biology of Machines, Social Systems and the Economic World* (1994), *New Rules for the New Economy* (1999), *What Technology Wants* (2010) |
| 1960– | Jaron Lanier American | Futurist, writer, composer, and computer scientist. Pioneer in the field of virtual reality. Key publications: *You Are Not a Gadget: A Manifesto* (2010), *Who Owns the Future?* (2013) |
| 1961– | Andrey Korotayev Russian | Anthropologist, sociologist, and economic historian. Prolific writer and pioneer in fields such as world systems theory, cross-cultural studies, and mathematical modeling of social and economic dynamics and cliodynamics |
| 1964– | Clay Shirky American | Professor and writer. Focus on socioeconomic and cultural effects of Internet on society. Key publications: *Here Comes Everybody: The Power of Organizing Without Organizations* (2008), *Cognitive Surplus: Creativity and Generosity in a Connected Age* (2010) |
| 1970– | George Dvorsky Canadian | Futurist, bioethicist, and transhumanist. Ethical and sociological impacts of emerging technology, specifically, "human enhancement" technologies |

*Source:* Kjaer Global

objectives. One of the first books ever written on future studies, *The Language of Forecasting,* with a French-English vocabulary by François Hetman, was first published in 1969.[2] The foreword notes that forecasting, as developed by Gaston Berger, originally focused on the philosophical questions, later developing an economic twist from Bertrand de Jouvenel and then adding politics and sociology to the mix. It adds:

> In view of Society's accelerated technological progress however, it is normal that the work carried out in this field, particularly in the United States, should develop the fastest.[2]

Certainly, this perspective proved to be correct, with many of the early trend forecasting methods developing and evolving within the US. However, what strikes me most about this early evaluation of an emerging field is the succinct and utterly persuasive vindication for its existence.

We see a clear recognition that the science of forecasting is about much more than divining how the future *might* turn out – it enables us to *influence it* and adapt to society's challenges by making sounder decisions about the future in the present. The book adds what I believe is a salutary reminder that the business of futures has a higher purpose than simply predicting ways to boost market share:

> The future is therefore our most precious resource. Its methodological exploration becomes a new dimension of our society. Concern for its implications must therefore increase rapidly.[2]

Discussion continues to this day about the best systems and methods – and even the best language – to decode the future. The current exponential data flow and our increasingly interconnected society adds to this debate, as we become ever more information burdened with shorter lead times. The principal question is: How do we evolve a framework that balances the tangible scientific and social data insights while also recognizing the impact of the intangible, value-driven changes in our society?

There are many examples of recent events – distrust in economic models and corporate governance, technology-driven behaviors, and grassroots movements for democracy, to name just a few – that simply couldn't have been foreseen, let alone explained, by reliance on pure data alone. So, the biggest challenge in the 21st century is to use a model that is adequate to validate and satisfy theoretical demands – scientific research and statistical data – while also considering cultural ecosystems that are the human, emotional aspects of the future. As *The Language of Forecasting* puts it, forecasting "must assimilate the material which is the subject matter of its activity, be it logical, exploratory or normative."[2]

Whatever methods are used, we must consider the core purpose of trend forecasting in order to benefit from the process. To make it meaningful, it needs to encompass an analytic framework, as well as an intuitive vision of possible events. With that principle in place, we are much more aware of how to actively realize our vision of a better tomorrow.

SUMMARY: Trend management

- At its simplest, trend management observes specific changes or developments and considers the general direction in which society is moving.
- Trend mapping and scenario building are part of the process and build a model applicable to brand building, communication, innovation, and future business strategies.
- The Trend Atlas is used to visualize research findings and provide a framework that enables the creation of future narratives.
- The ultimate goal of trend management is to view our organization holistically, considering our current position and future choices within a society-wide context.

## What this book can do for you

Working out how to best navigate the future is now a global debate with many crucial questions to be answered. What leadership qualities are needed and how do we prioritize now and down the line? Do we opt for a more human-centric and inclusive society model or adjust current structures and let the market take care of things? Conversations have always taken place at every level, from government through to business leaders, media and citizens, so this is not a new discussion. The real game-changer in recent years, however, is the channels used, with the Internet enabling real-time dialogue and placing people, not governments and businesses, in charge of directing the conversation and shaping expectations about tomorrow's world.

As an organization anticipating the future, you have two choices. You either sit back and wait for the debate to reach a firm conclusion and change to take its course or you become an influencer – acting on the future before it acts on you. Clearly, in my line of business, I advocate becoming an active change-maker as the only forward-looking strategy. To be a successful organization or – come to that – individual in the 21st century, we must learn how to approach change with confidence by recognizing that current and future shifts also represent some of the greatest opportunities for fresh ideas, growth, and success.

True confidence comes from anticipating likely outcomes – including discovering a need or gap in the market before your competitors – and that means organizations must engage in honest and open debates with all stakeholders to build a foundation of future intelligence. Stakeholder debates don't provide all the answers, but they do raise the crucial questions that need to be addressed within strategy in order to formulate a viable direction for your organization.

It is already widely accepted that gathering and analyzing information is a necessity for any business wanting to stay ahead in the knowledge economy. Companies make future mapping – in which they extrapolate likely developments in their markets based on current data – an integral part of their business practice as a means of becoming more innovative and competitive in an ever-changing landscape. It has huge potential and benefits for any organization that is serious about being ahead because success and growth depend on the knowledge, creativity, and commitment of the people working for you. Organizations also have to cultivate a culture of innovative thinking based on behavioral economics – monitoring current trends in society to understand people and their behavior, needs, and wants. And this is where trend management comes into its own because it starts with a systematic process of trend tracking.

Monitoring and analyzing local and global society drivers by extrapolating forward to recognize developments in the medium- to long-term future is what trend management can do, but its benefits to your organization extend far beyond that to supporting and shaping organizational mindsets and culture. This is because the shared act of considering where your organization is now and where it wants to be in the future not only takes the conjecture out of the process, but also acts as a unifying force that boosts people's shared sense of working towards a common goal. And this is exactly what this book prepares you for: visualizing and navigating the change-making process. The Trend Management Toolkit system brings you closer to people by giving you profound insights into the trends that influence and shape all of us – in society, in business, and as citizens. By understanding that we are all part of a vast global network in an intertwined system, we start to influence and shape our future.

## Creating your map of the future

Using trend management tools can teach people and organizations how best to grow together and make positive change happen in a directed and organic way. As a futurist, my business is not to present all the answers, but rather to act as a guide by sharing inspirational thoughts and ideas and then inviting companies to map out their challenges, strengths, and dream scenarios. In this way, they are able to assemble a future road map to clearly define the journey ahead. Evaluation tools such as a SWOT (strengths, weaknesses, opportunities, threats) analysis assist with this. We also employ our Trend SWOT analysis to create an individual "map of experiences," as this helps companies identify the opportunities and threats they face, and then consider how to capitalize on their strengths and use them to address their weaknesses.

This is a practical step-by-step process that begins with research into contexts – society, culture, and people – to present a bigger picture in the form of a Trend Atlas in conjunction with carefully studying a company's culture, market, and operating environment. As part of this process, we monitor and analyze emerging trends to extrapolate how the market might look for the organization in the medium to long term. This is a process that evolves as inspirational ideas are discovered and developed, generally undergoing several iterations before a final plan emerges. The trends we incorporate and discuss provide insights into socioeconomic and cultural drivers, opening a window into the demands and needs of tomorrow's people. Within this process, we recognize that there are huge opportunities in supporting people – internal and external stakeholders – in making their lives more meaningful, and also in inviting them to take responsible action. This means our vision always works from the point of view of wanting to improve people's situation and current lifestyle.

Placing people at the center of your thinking and innovation processes makes good business sense and, in this context, it is important to point out that the concept of "stakeholders" has evolved to include a much wider ecosystem. It is no longer restricted to staff and investors, but also wider communities of customers and fans who participate in shaping brand values and individuals who choose to contribute as free agents of change. These free agents may even extend to social networks of influencers, critics of your brand, and other grassroots activists. We consider this much wider definition of stakeholder essential because, over the past decade, social networks have profoundly impacted not only the brand and business landscape, but all sectors of society. This is undoubtedly a growing force – from the Obama effect and the Arab Spring to radical transparency exemplified by WikiLeaks and Edward Snowden – and with a ripple effect that cannot be ignored in future organizational planning.

## Are you ready for change?

Being future oriented requires us to actively allocate time to think about change and how it could impact our operation. A positive commitment to be proactive rather than reactive is essential since most of us are generally too busy with day-to-day operations to plan ahead. Change agents are individuals who make change happen and they are typically resilient, self-reliant, and extremely potent professionals who work solo, in small teams, or within large companies – every organization, large or small, needs them to thrive. We can all become change agents and the first essential step is to break old patterns of behavior and recognize the value of taking time out to think about the future.

So how do people justify the required time investment? There is no easy answer, but if fear – of the market, the opposition, the balance sheet – rather than positive anticipation rules your organization's everyday agenda, then it needs a change agent. Often, entrepreneurs work on their businesses for long hours out of love for what they do and that is fine as long as it is a motivator, but doing it to clear the paperwork mountain and meet deadlines is not a justification. And, as an individual, if your output and your innovative ideas are drying up in the face of eternal firefighting and you only have the opportunity to consider things through the prism of the next set of targets, then it's definitely time to step away from the office to reignite your passion for what you do.

The value of enabling this process has already been recognized by some organizations. Microsoft, Apple, and Google employ intrapreneurs to stay innovative and be in the loop – they have made cultivating change a part of their culture. Google's "20 percent time" program enables key staff to take specified time (one-fifth) out to work on individual projects that aren't part of their day-to-day role in order to help develop something new or improve existing thinking. This initiative actively invites more than just top leaders to take control over the future, it must be encouraged on an individual and organizational level.

There is a major cultural and mindset change required for some organizations in order to begin the process of opening the door to future planning. They need to understand that time out to think is not time off, but rather an active process of capitalizing on change. In fact, pioneers of this was 3M who launched the "15 percent program" in 1948 as a logical next step after the first early years in the red. It was a radical step back then where rigid organizational hierarchies defined work – and taking time off to think was radically different – but it paid off, with the Post-it note being a result coming out of that time. 3M's ethos of "innovate or die" has been carried into the 21st century and the 15 percent time to think program is now key to 3M's business strategy.[3]

The action of considering and then possibly breaking away from routine patterns of thought, planning, and established systems is hard work – actually much harder than sitting in the office undertaking day-to-day tasks. It's stimulating, sometimes frightening, but also immensely rewarding as an experience of process. All the justifications in the world won't kick-start the change process until you and your organization are ready. But the fact that you are reading this book suggests that you are at least contemplating and preparing for your journey into the future.

## Trend versus fad

We all like to know we are on track and making informed choices, but how do we distinguish between a trend that will develop into the future, and a fad that will be gone by next year? There is no easy answer to this, and this is why companies turn to forecasters. In recent years, we have seen so-called "cool hunters" and "trend hunters" emerging onto the scene – promising to find us the next big thing.

Today's forecasting industry is so diverse that it can be hard to find out who can best help you with your particular business challenges. Forecasters across a diverse range of industries have very different approaches. Success rates vary and this may be down to the individual forecaster, the brief or simply because the area of strategy or the organization's expectations are not clearly defined at the start of the process. Forecasting and then implementing future directions is most likely to succeed when organizations set out clearly defined objectives at the beginning by inviting their core players to participate in the process, mapping the challenges ahead, and the potential rewards of solving them. This gives the process goals and meaningful objectives that can be conveyed to all stakeholders. Trend management is a system that enables companies to stay in control by helping them separate the passing fads from the meaningful movements and shifts that look likely to develop into more influential trend drivers. By subjecting trends and new ideas to close scrutiny in a process of collaboration, there are far fewer opportunities to miss the obvious and to successfully utilize research as a means to verify the strategies you are thinking of pursuing and acting on.

Like an author or filmmaker, the futurist is a professional narrator and storyteller, whose main objective is to observe key sources and events and then connect all the dots to visualize their impact and outline the bigger picture. But bringing future stories and strategies to life is essentially a participatory exercise – which is why the process is best undertaken within a structured trend management environment in which all players recognize the vital role they play.

## Understanding trend forecasting

There are distinct strands of future studies and strategic management that have informed and shaped the current landscape of forecasting. We describe them as the scientific, social, emotional, and wild card approaches (see Figure 1.1 at the end of this section). While it is important to understand each individual strand's history and context, the approaches and methodologies described here are not entirely independent of one another and no single method or framework provides all the answers. For that reason, the Trend Management Toolkit supports a balanced combination of methods to explore the "bigger picture" before narrowcasting into the most relevant areas that represent future challenges and opportunities for an individual organization. Our multidimensional forecasting approach – the core of the Trend Management Toolkit system – is discussed in more depth at the conclusion of this section.

### Scientific forecasting

The most commonly known system is "scientific forecasting." This method makes predictions based on statistical (usually regression and time series analysis) and mathematical models built on sources such as the World Bank, the OECD, National Statistics, the UN, and NASA. Traditionally, the resulting forecasts were employed within government and large organization research teams to understand complexity and foresee tipping points. Today, most public institutions and a wide range of private organizations subscribe to reports or produce their own medium- to long-term forecasts using quantifiable scientific data.

Originally, this work was not classified as trend forecasting – instead it was termed "prognoses." But then this was in an era when the rate of change was gradual and you could place more faith in assessments of likely future outcomes based on past markets and structural patterns. Historically, biology and physics produced the best statisticians, as it was crucial for the development of those fields. Since the 1960s, there has been a vast emergence of economic and political analysts dedicated to understanding the mechanisms of financial markets and the business landscape, usually based on mathematical formulae, and the finance and industrial sectors still favor scientific forecasting's linear approach to profit maximization and economic growth.

Scientific forecasters are effectively visionary number crunchers who correlate and analyze macroeconomic indicators, demographics, and public policy. The best-known practitioners are academics and analysts operating from a facts-based point of view. They look for relevant numbers to extrapolate how the future may unfold, often related to growth versus decline in a particular sector. Alongside data from the financial sector, fundamental forecast tools typically include impact assessment and scanning activities related to technology and life sciences developments.

### Social forecasting

This system for predicting lifestyle and consumer behaviors is the most extensive of them all – a method I categorize as "social forecasting" – and is based around the study of the development of social groups and their dynamics. Researchers, most of them drawn from the fields of sociology, anthropology, ethnography, and psychology, use specific methodologies to study human behavior, cultures, global communities, and local characteristics. This approach deals primarily with human interactions and assumes bounded rationality as decisive to behavior and profit. Findings are used to understand the value-driven side of human behavior.

In the early decades of the 20th century, social psychology was employed to study human behavior for the purposes of understanding motivators for engineering consent and tapping into people's unconscious desires – all for the purposes of increasing sales and developing new markets. It proliferated during a time when the concept of the consumer society was being shaped and traditional market research was no longer considered an adequate tool to understand and manage the dramatically increasing demand for goods and services and the changing social status among an upwardly mobile working population.

This group of forecasters uses multiple methods of empirical research and critical analysis, combining it with interaction-based, expert interviews to develop and back up a trend framework at macro and micro levels. A wide range of content is studied, ranging from cultural habits in society and family roles to the environmental, political, and economic impact of lifestyle patterns. Over time, hermeneutics, semiotics, and philosophical approaches have been applied to the analysis of society. More recent additions to the discipline include action and interaction-based techniques, such as agent-based modeling and social network analysis.

### Emotional forecasting

The field of what is best described as "emotional forecasting" emerged and evolved from the design and creative industries. This trend-watching system is used to predict developments within fast-moving fields, such as clothing, food, interiors, paints, cosmetics, retail, media, and marketing – areas traditionally influenced by fashion and cultural shifts. This approach is also extensively used by the automotive and technology sectors, where CMF (color, material, and finishes) strategies are developed to create an emotional consumer bond – in other words, make us fall in love with the product. A typical example of this is in the field of mobile technologies (phones, computers, tablets), vastly crowded arenas where the design packaging of the product has become almost as essential as functionality. These forecasting communities are masters of creation and re-creation, in which design, material, and color play a key strategic role in developing an engaging consumer experience.

Typically, forecasters work in small firms as consultants or advisors, or as part of a specialist in-house team. Many of the most influential started out in Paris, London, and Milan in the 1970s, later spreading out to other key urban centers. Today, it is not enough to identify customer demographics, companies also have to study the psychographics of attitude, beliefs, mood, values, and situation. This methodology deciphers emerging consumption, social, cultural, and aesthetic markers, translating them into actionable creative concepts in local and global markets. Emotional forecasting is an important means of ensuring that new products and services capture the zeitgeist, referencing the relevant factors affecting consumer moods and choices. These forecasters are often design trendsetters, guided by intuition and a powerful sense of the "next big thing."

One of the driving forces behind emotional forecasting is to stimulate the consumer to "need" a product – often a discretionary spend – and the system of predicting what consumers will want necessitates a balancing act between the anticipation of future developments and improvisation. In this regard, this is a field of market influencers and innovators; notably, many forecasters are subject experts in color or design but they also keep an eye on wider cultural and arts movements, as well as science and technology developments. This discipline adds a deeper emotional layer to the product planning and innovation process. It is also a valuable support tool for R&D in reaching existing or new target markets with on-trend or ahead-of-trend products and services.

### Wild card forecasting

Wild card forecasting is distinctly different from the other systems, and certainly not yet part of the mainstream industry; however, it is having an important impact on the overall shape of the future trends landscape and cannot be overlooked. Individual experts or small think tanks create forecasts and analysis based around scenarios of low probability, but with high impact. Typically, the forecasting mechanic incorporates quantitative and qualitative data to come as close as possible to a scientific analysis. One area in which this aspect of forecasting is proliferating is within business literature; notably, Nassim Nicholas Taleb's *Fooled by Randomness: The Hidden Role of Chance in Life and in the Markets*[4] and *The Black Swan: The Impact of the Highly Improbable*[5] have opened up debate into unexpected events and increased interest in how they may become game-changers. The black swan of the second title is a motif for the financial shifts, pandemics, and major historical events that may be considered as undirected and unpredictable.

While the analysis metrics are highly speculative in nature, I consider it valuable to incorporate wild card exercises within forecasting as they encourage us to imagine the "unthinkable." In these exercises, we should remind ourselves that, while there are always events outside our control, we also have enormous influence because the future strategies we create have the power to shape better business, social, and environmental ecosystems.

It should be added that wild card forecasting often focuses on a specific reference frame and viewpoint – notably risk ideation and management and the use of "what if?" thinking – but recognizes that in every risk also lies opportunity. Such forecasters commonly operate to support strategists, project and risk managers to help them identify and define events that might disrupt and impact current systems – typically from a political, economic or environmental point of view – and utilize specialist data and formulae alongside creative thinking to plan for adversity and balance possible risk with potential rewards.

## APPROACHES OF FORECASTING METHODS

**Scientific:** Evidence-led and analytic – methods typically aim to forecast a particular field supported by reliable documentation data analysis, statistics, and quantifiable indicators.

**Social:** Interactive and analytic – uses "bottom-up" and participatory processes to balance expert view with non-expert stakeholders, as well as evidence-based structural analysis.

**Emotional:** Intuitive and visionary – methods that blend media scanning, culture, art and design inspiration with input from industry "gurus," storytelling, and brainstorming sessions.

**Wild card:** Expertise-led and visionary – methods use the skills and expertise of individuals in a particular area such as science fiction to provide advice and make recommendations.

## Multidimensional forecasting

All the single-strand forecasting systems currently being practiced have their qualities but also their limitations, and I believe that the complexity of the 21st century calls for a balanced approach that is also flexible enough to respond to the contexts and challenges across business sectors and organizations. While we incorporate elements of various approaches, I like to emphasize that any forecasting exercise is a positive one, in that our actions and choices actively shape our world and have the power to influence final outcomes. Therefore, I view it as essential to look at the bigger picture and explore the dimensional aspects of trends – what we refer to as "whole-brain thinking." For this reason, our trend management system is overtly interdisciplinary and explores the scientific, social, and emotional dimensions, also considering elements of wild card scenarios.

The various forecasting methods may be used at different stages in a trend-mapping process – depending on the business area and time horizon. This takes foresight to the next level, as a systematic, participatory, and visionary process that supports businesses to create their strategic future road map. The Trend Management Toolkit actively invites key stakeholders to engage in a guided and structured discourse about possible futures in order to create a common platform for a deeper understanding of which long-term issues to anticipate and plan for.

### TREND RESEARCH: MOST COMMON METHODS

**Literature review:** Part of the trend-scanning process – generally using a discursive writing style structured around themes and related theories.

**Expert panels:** Groups of people dedicated to discussion and analysis, combining their expert knowledge or given interest area.

**Scenarios:** Using a wide range of methods and data insights to build plausible future scenarios based on systematic visions of how the future could unfold.

**Workshops:** Typically consist of a mix of talks, presentations, discussions, and debates on a particular subject, where participants are assigned specific detailed tasks.

**Brainstorming:** This interactive method aims to develop out-of-the-box thinking in creative face-to-face sessions to generate new ideas around a specific area of interest.

**Trend extrapolation:** The most established tool and forecasting method to provide an indication of how past and present developments may evolve in the future.

**Delphi:** Repeated polling of the same individuals, to build a PESTEL trend argument by allowing for best and open judgments in a feedback and analysis process.

**SWOT:** Method identifies internal strengths and weaknesses to tap into potential and prepare for external factors in terms of opportunities and threats.

**Interviews:** A fundamental social research tool – used within a consultation process to gather knowledge across the range of interviewees.

**Monitoring:** Use of tools such as Google Analytics, instant feedback, and peer reviews gathered via the Internet.

The Trend Management Toolkit draws widely from a range of forecasting methods that are mapped out in key forecasting methodologies (Figure 1.1). According to a 2009 EU study *Mapping Foresight: Revealing how Europe and other World Regions Navigate into the Future,*[6] the most commonly used forecasting methods across all industries are literature review (54%), expert panels (50%) and scenarios (42%), workshops (24%), brainstorming (19%), interviews (17%), surveys (15%), scanning (14%), SWOT (11%) – all of which are qualitative. Other methods include trend extrapolation (25%), which is quantitative, plus key technologies (15%) and Delphi (15%), both semi-quantitative.

## SUMMARY: What this book can do for you

- Today, people are in charge of the debate about our shared future, so organizations must engage in honest debate with all stakeholders to formulate a viable future direction.
- Note that in today's brand universe, the concept of stakeholders has expanded to include fans, critics, and free agents of change, such as social networks.
- Trend tracking fosters innovation and shifts organizational mindsets and culture by acting as a unifying force so that people can understand and share common goals.

- Scientific, social, and emotional forecasting build a holistic picture of how the future may evolve, while wild card forecasting considers disruptive and unexpected events.
- The "whole-brain" or multidimensional approach to forecasting combines the strengths of each method to reveal micro influences that make up the macro trends and key society drivers.
- The final objective is to design a framework that works across every level of an organization, culminating in visionary multidimensional thinking and practical future strategies.

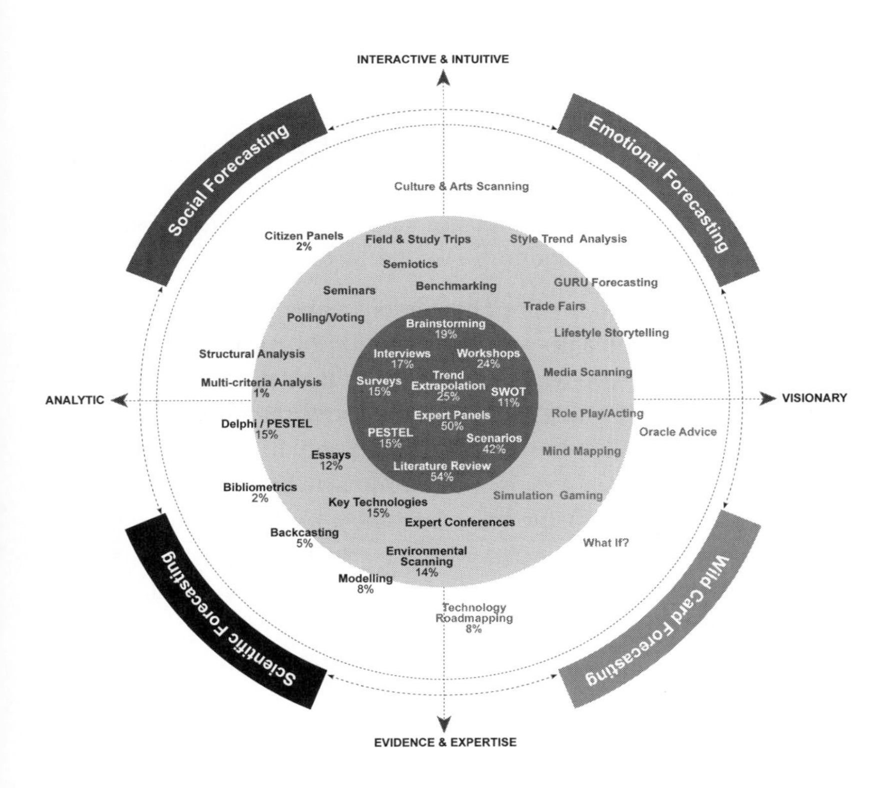


FIGURE 1.1 **Key forecasting methodologies**: 4D model of key forecasting types with multidimensional forecasting in the center
*Source*: Kjaer Global

## Taking time out to think and act

So why am I such an advocate for taking time out to think and envision the future? Well, in order to influence markets and connect with your audiences, you need "big dreams" because these are what win you loyalty and market share. As Walt Disney pointed out: "If you can dream it, you can do it."

However, when we look at today's corporate culture there is still a long way to go for the majority, as most thinking revolves around calculated risk and projected profit. In my early years in the forecasting industry, an artist friend of mine pointed out that thinking is often not really valued in our society because, as he put it: "Thinking is seen as nonproductive – people want tangible output that can be measured and turned into a sale." When we contrast this attitude to the reverence we feel for the great philosophers and scientific thinkers of history, this does seem a surprisingly short-sighted approach because they have profoundly influenced and directed the cultures and truths we live by today.

Having often reflected on this observation, I find that now, almost 20 years later, more companies are finally starting to engage in the kind of deep thought and vision required to work out their role and contribution to society – they are gradually joining the conversation. This, for me, is a huge positive as it shows corporate recognition of the need for meaningful contemplation in a much wider context – balancing the tangible calculative thinking with the more intangible meditative thinking as a means to produce better results. I believe that this new engagement is a sign of our times, in that we are moving towards a more inclusive and compassionate era.

### Allowing for the unexpected

It can be hard to trust in the process of balanced thinking as a method to find inspiration and practical direction, as we often tend to subscribe to one school of thought – with a preference for either pragmatic or intuitive thinking. Certainly, to maximize the process, it is important to recognize that we may not find the answer we were expecting. In science, this is a well-established process, as many discoveries, from penicillin to Viagra, have been made on the road to searching for another application, while drugs initially assumed to be abject failures have turned out to have life-saving applications elsewhere.

Many corporations are still not comfortable with "thinking time" and ideation as active tools to move forward, so one way of rationalizing their importance is to view them as future capital – a vital part of a sound planning process. In 2003, we started a series of conferences called Time to Think, where we invited people from a broad range of industries to take time out to think with us. Our friends from the Copenhagen Institute of Future Studies[7] – known for the book *The Dream Society: How the Coming Shift from Information to Imagination Will Transform Your Business*[8] – joined us as partners. It was an inspiring and fruitful experience that made us see the world around us in an entirely different light. I now fully trust the process of thinking time and know that this is the best fuel for innovation and positive change. With this personal experience behind us, when we start a future project with a client, the destination is never set in stone – openness to the unexpected is a requirement of the process, signaling a willingness to bring on real change by engaging in an honest exploration where we mutually evolve with the project, the organization, and the people.

## Balancing analysis with informed intuition

So where does analysis stop and innovation begin? Traditional linear thinking can often become a stumbling block for generating fresh ideas – hence it is important to nurture the unique and the radically different, disruptive approaches in order to create a strong identity and a solid, future-proofed direction.

The world of today still mostly engages in left-brain processes, only considering factual thinking and everything that can be measured. However, while statistics reveal past and present patterns, they seldom tell us much about the future. This is why, over many years, we have developed and proposed a system that integrates right-brain thinking; it delivers a clear focus on qualities of intuition and vision, thus providing essential balance in the forecasting process. By visualizing the whole picture and considering society's currents and how these relate to people's real needs and wants, we get closer to a human-centric outlook. This process necessarily means we step outside our corporate box in order to envision how our organization appears to others – using whole-brain thinking. Because it provides a broader and more holistic view of our surroundings, it becomes far more likely we will innovate in a way that appeals to our audience.

Edward de Bono,[9] the father of lateral thinking, described in 1967 how thinking outside the box is a skill everyone can learn, but requires an alternative approach to unleash people's creativity. While I don't necessarily follow de Bono's core philosophies, I support the principle that creativity is not a gift, but a process that can be nurtured by using a set of structured arenas where it is enabled and encouraged.

## The ongoing process of navigating complexity

Globally, the business landscape has transformed immensely over the past 50 years. As part of this change, we have witnessed a significant shift from a ME approach involving a few influential stakeholders to a WE approach, in which large communities of people are in constant dialogue about our society, business values, and ecosystem. Currently, most organizations still use rational approaches alone to navigate complexity, but what we need are models that are agile, scalable, and intuitive enough to explore the unknown.

The American futurist Herman Kahn planted the early seeds of scenario planning. While working as a defence analyst at the Rand Corporation in the late 1940s, he started telling brief stories to describe possible impacts of a nuclear war and ways to improve survivability – he was imagining the worst of all possible futures by mapping out scenarios. The pioneering French oil executive Pierre Wack, who later invited Kahn to help him adapt and further develop the use of scenario planning for Shell in the 1970s, said this process is about "being in the right state of focus to put your finger unerringly on the key facts or insights that unlock or open understanding."[10]

Their method was all about insight, complexity, and subtlety, not just formal analysis and numbers. Wack had a brilliant story about how intuition works. A Japanese gardener had once told him that if you throw a pebble at a thick bamboo stem and it hits the trunk slightly off-center, it will bounce off, hardly making any sound. But when you hit the trunk dead center, it makes a distinctive "thump" sound.[10] This was how Wack described the process of intuition, saying that you must know the sound you want to hear in your own mind before you even throw the pebble and then listen out for it.

Intuition may not be the reassuring sound of a pebble on bamboo stem for all of us, but Wack was clearly describing a process of learning to trust our judgment; and it's worth remembering that he was prodigiously successful at foreseeing forthcoming oil crises and, in an era when the petrochemical giants were struggling to change course, he proposed a new model of flexibility and awareness. There is space beyond the known world and its hard data sets that we must become familiar with and comfortable inhabiting if we want to plot the future with confidence. The results, as described by Wack in Scenarios: Uncharted Waters Ahead, a 1985 *Harvard Business Review* article,[11] make the effort worth it:

> This transformation process is not trivial – more often than not it does not happen. When it works, it is a creative experience that generates a heartfelt "Aha" … and leads to strategic insights beyond the mind's reach.

## INTUITIVE DECISION MAKERS

In the 1980s, John Kotter studied the behavior of a group of top executives and found that they spent most of their time developing and working a network of relationships that provided general insights and specific details for their strategic decisions. They tended to use mental road maps rather than systematic and structured planning techniques. In 1984, this point of view was backed up in a study by Daniel Isenberg, which concluded that senior managers make highly intuitive decisions.

Companies that actively engage in the process of change were described as "learning organizations" by Arie de Geus in the 1990s and this concept was later popularized by Peter Senge[12] *in The Fifth Discipline: The Art and Practice of the Learning Organization.* Both de Geus[13] and Senge were describing environments in which gathering and analysing information in order to evaluate and form fresh ideas and thinking patterns were part of the culture. The learning organization encourages its people to see the bigger picture together and act on it in order to perpetuate the organization. Such an organization is an organic entity capable of learning and creating its own processes, personality, and purpose.

Due to the rising complexity in the world, organizations have to explore how to best adapt to change. A common feature of companies that have prospered for at least 50 years is that they have adopted an intuitive learning model to further their innovation process and business survival strategy. In most cases,

> it has not necessarily been an even curve of constant progress but a bumpy journey of lessons to be learned.
>
> IBM is a good example of an organization founded on old-style business principles that found instilling creativity throughout its organization was the only route to navigate change. The tech and office giant was facing a multi-pronged assault: industry transformation, shifts in global economic power centers, new legislation, increased data volumes, and new customer preferences. The utilization of online discussions with its internal stakeholders (staff) in 2003–4 led to a rewriting of company values and, ultimately, a redefinition of company direction.

In today's business environment, we know that an overreliance on any single strategic approach would not be wise. Organizations need multiple methods that combine intuition and analysis, as this is fundamental to creating an approach to the future that cannot easily be replicated by the competition. Intuitive strategic management is an important part of cultivating the future as well as engaging in an ongoing process that keeps a keen eye on your business direction and the environment in which you operate.

Navigating a steady course is never easy, but Wack's visionary 1960's approach does resonate with the times in which we live. To borrow Wack's pebble on bamboo metaphor, the most important question for any 21st-century organization to ask is: What is the sound that most resonates with my future? Then you can tune into that sound for signs that point you in the right direction. After all, the future is not just somewhere we go to, we create the future.

## SUMMARY: Time to think

- Corporate culture still focuses on projected profit, but the process of future planning builds performance by inviting "big dreams" to win loyalty and market share.
- With trend management, your final destination is not set in stone, rather it is important to nurture disruptive and radically different ideas within the step process.
- "Whole-brain thinking" overcomes traditional overreliance on left-brain (facts-based) thinking to nurture more intuitive and creative approaches.

- Taking ownership is vital to successful innovation because when we feel personally involved we become influencers and active change-makers.
- Future planning must recognize the shift from a ME to a WE focus in society, in which people-centric products and services will be required.
- We must develop multidimensional strategies to produce agile, scalable innovations and business models that cannot easily be replicated by the competition.

# Chapter 2

# Sense making in a fast-forward society

*Most of us do not feel happier or more empowered by the multiple choices on offer in a speed obsessed world. Instead, we feel overwhelmed and want to switch off.*

One of the main challenges in the Western world is that of overconsumption. To consume or not consume has become our big ethical dilemma – one that seems to unite us, but at the same time divide the haves and have-nots. And when marketers talk about meaningful consumption, they focus on a small proportion of people; according to the World Bank, 80% of humanity lived on less than $10 a day in 2008.[1] In the *Report by the Commission on the Measurement of Economic Performance and Social Progress*, Joseph Stiglitz, Amartya Sen, and Jean-Paul Fitoussi observed: "Those attempting to guide the economy and our societies are like pilots trying to steering [sic] a course without a reliable compass."[2] I suggest that what is missing from our current navigation charts is a moral compass and this will not emerge until we address the root causes of inequality at a global level.

Meanwhile, we consume 26 times more than we did 150 years ago.[3] It is no wonder that the word "consumer" has taken on increasingly negative connotations, and many find the term off-putting, even derogatory. John Mackey and Raj Sisodia, in their book *Conscious Capitalism*,[4] argue that caring businesses see their customers as human beings whom they are privileged to serve, rather than "consumers to be sold to." They add that: "the very word 'consumer' objectifies people, suggesting that their only role is to consume." Groundbreaking books, such as Naomi Klein's *No Logo,*[5] published just before the dawn of the new millennium, fueled the current anti-consumerism movement, which rejects the obsession with brands and conspicuous consumption. Of course, the act of consuming is not always guilt-ridden; many enriching and life-enhancing experiences also spring from our choices, so it is not all bad news. But maybe it is time to replace the term "consumer" with "people," to connect us back to an authentic landscape of real human needs and desires.

As the global dialogue around consumption intensifies, people begin to ask themselves profound questions. This new awareness is impacting behavior and lifestyle choices, presenting huge opportunities for businesses in helping individuals achieve higher levels of meaning through what I call "inspired engagement." Increasingly we ask: How can I get more from my life?, and it is important for brands to connect with people in a way that enriches lives, not just company bank balances. Tania Singer, director of social neuroscience at the Max Planck Institute for Human Cognitive and Brain Sciences, addressed this question in Beyond Homo Economicus:[6]

> Humans are capable of far more than selfishness and materialism. Indeed, we are capable of building sustainable, equitable, and caring political systems, economies, and societies. Rather than continuing to indulge the most destructive drivers of human behaviour, global leaders should work to develop systems that encourage individuals to meet their full socio-emotional and cognitive potentials – and, thus, to create a world in which we all want to live.

## Connecting the dots

The Trend Management Toolkit is a system developed for observing society trends and their impact on human behavior. Translating these findings into viable innovation concepts for a sustainable business vision is the exercise of connecting the dots to make sense out of what seem complex and, at times, contradictory elements. But we can only connect the dots we have already spotted – hence, observing trends requires intuition and good judgment in order to see the bigger picture. In effect, these dots are the key to how we will visualize the future (Figure 2.1).

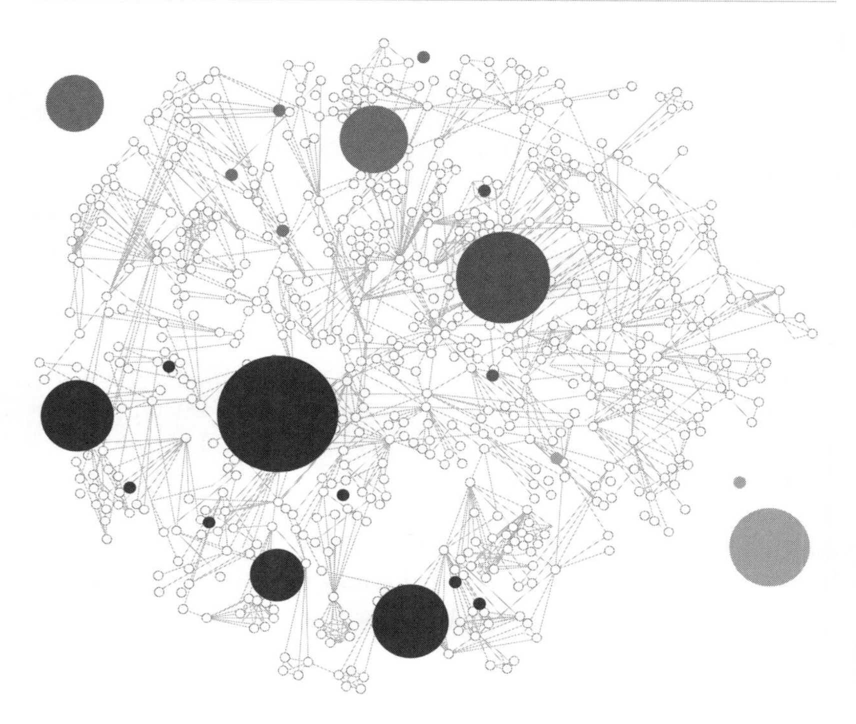

FIGURE 2.1 **Mapping the trends**: Spotting the right dots is essential to see the bigger picture
*Source*: Kjaer Global

In his now famous 2005 Stanford Commencement Address,[7] delivered to graduating students, Steve Jobs noted:

> You can't connect the dots looking forward, you can only connect them looking backwards. So you have to trust that the dots will somehow connect in your future. You have to trust in something: your gut, destiny, life, karma, whatever. Because believing that the dots will connect down the road will give you the confidence to follow your heart, even when it leads you off the well-worn path.

## Understanding the evolution of consumption

Looking at the evolution of needs over the past century, there is an apparent development in the way we view the world – a big cognitive shift where consumer patterns and sociopolitical models have emerged beyond the 20th-century trajectory. Knowing that material wealth alone is not the route to happiness, we are discovering different paths for personal empowerment. The evolution of consumption diagram (Figure 2.2), inspired by Abraham Maslow's 1943 hierarchy of needs pyramid,[8] illustrates the history of our society and the rise of consumerism. Maslow sought to explain human "progress" beyond basic needs and looked to psychology to discover the motivation for human "betterness." Our evolved diagram describes the ongoing relationship between the individual and organizations since 1900. Showing the clear progression into a post-consumer age, in which we care less about ownership and more about access, Figure 2.2 also explains the evolution of what we have termed the "good life" society, in which the individual is motivated by meaningful brand engagement. We will explore these themes in greater detail, beginning with an exploration of the origins of the "good life" philosophy in the next section.

The consumer society debate is not a new one; Thorstein Veblen, the American economist and sociologist, coined the term "conspicuous consumption" in *The Theory of the Leisure Class* in 1899.[9] His treatise pointed out that, as societies become wealthier, leisure time increases and overt consumption becomes a means of expressing and displaying economic power and social status. Arguably, Veblen's observations can be seen today in the buying patterns of emergent and newly prosperous societies, most notably in the BRICA regions (Brazil, Russia, India, China, and the African continent), where a rising middle class is aspiring to more and better by buying homes, cars, travel, technology, and luxury items. Meanwhile, in the West we've transitioned to another place; we now start to view the reproduction of our own 20th-century spending patterns within developing economies with unease, but we are hardly in a position to criticize anyone else for wanting some of the possessions that fill our lives. Of course, in some regards, being a "conscious consumer" by spending thoughtfully to show you care is another form of status behavior – one that has been picked up swiftly by marketers – because often the socially and environmentally responsible choice comes at a price that puts it out of reach of many, even in the West.

The seeds of the current anti-consumerism movements were planted when the postwar intellectuals and thinkers came together in a perfect storm in the late 1960s and 70s and began to question our unstoppable economic trajectory, arguing that self-expression and quality of life should replace material status and become our primary goals. The debates continue as we redraw the map of how we want to live and more of us consider how we should achieve an inclusive economy – one where people and the planet are placed in the center of our moral compass. What is certain right now is that companies need to open up the conversation with their stakeholders and move beyond retrenchment or token gestures into an honest, two-way dialogue on how consumption patterns should evolve to support a sustainable and prosperous society that enables more people to enjoy the "good life."

## The economics of enough

In the documentary series *The Century of the Self*,[10] filmmaker Adam Curtis candidly portrayed the emergence of consumer culture, painting a ferocious but interesting historical portrait of the early seeds of consumerism. It showed how, in post-industrialization, companies manufactured to fulfill basic needs in a culture where people ate to sustain themselves and only bought the bare essentials. While the wealthy would buy things they didn't need, most people did not. But suddenly there was a shift in behavior and attitude – one in which people started buying things they essentially didn't need but wanted in order to enhance their sense of status and self-worth. The Austrian-American Edward Bernays – nephew of Sigmund Freud – was instrumental in fostering the American consumer society and changing corporate mindsets to serve new emerging needs. As a pioneer of public relations and propaganda, he used social psychology to understand and influence behavior.

While somewhat shocking when viewed through the prism of our modern sensibility and sensitivity to consumer manipulation, there was real human insight in Bernays' thinking and it worked like a dream, in the short term at least. By the early 1920s, New York's banks were funding the opening of department stores across the US to sell mass-manufactured goods. This era saw the creation of many of the persuasion techniques we still use today, such as product placement in articles, advertisements, and movies. Psychologists were employed to issue reports – typically under the guise that they were independent studies – claiming benefits to health, attractiveness, and success. Most of these early "scientific" claims may subsequently have been disproved, but the potency of the message

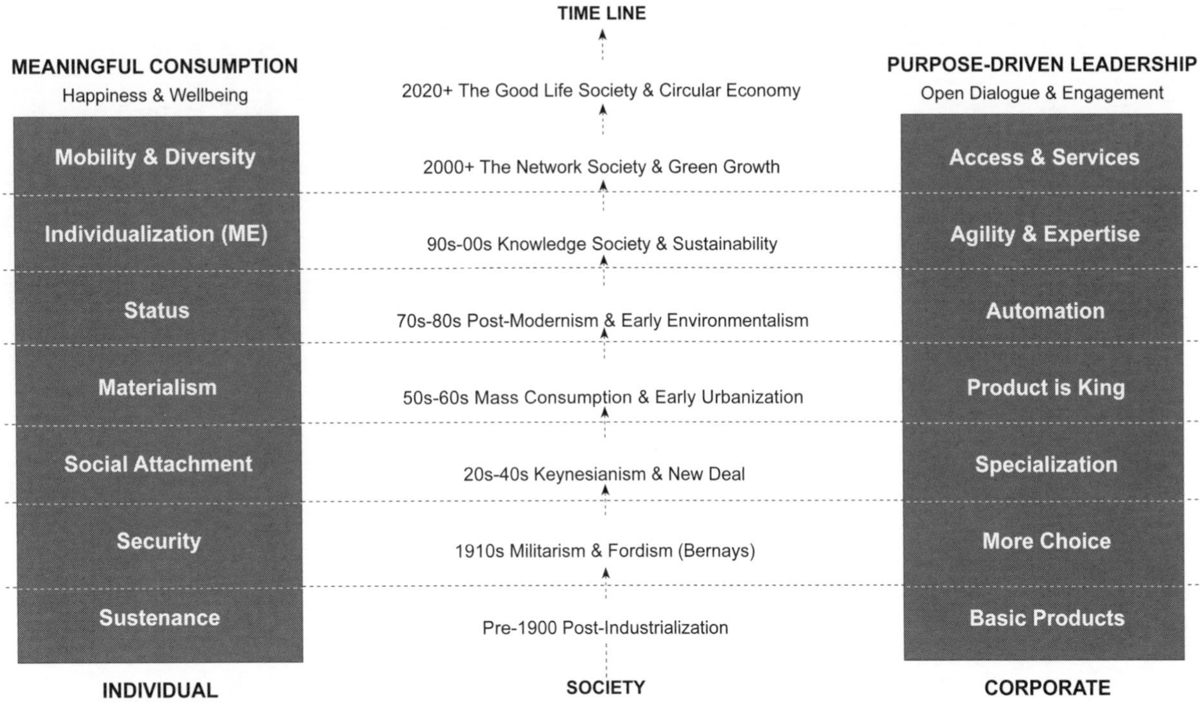

TIME LINE

| MEANINGFUL CONSUMPTION<br>Happiness & Wellbeing | | PURPOSE-DRIVEN LEADERSHIP<br>Open Dialogue & Engagement |
|---|---|---|
| | 2020+ The Good Life Society & Circular Economy | |
| Mobility & Diversity | 2000+ The Network Society & Green Growth | Access & Services |
| Individualization (ME) | 90s-00s Knowledge Society & Sustainability | Agility & Expertise |
| Status | 70s-80s Post-Modernism & Early Environmentalism | Automation |
| Materialism | 50s-60s Mass Consumption & Early Urbanization | Product is King |
| Social Attachment | 20s-40s Keynesianism & New Deal | Specialization |
| Security | 1910s Militarism & Fordism (Bernays) | More Choice |
| Sustenance | Pre-1900 Post-Industrialization | Basic Products |
| INDIVIDUAL | SOCIETY | CORPORATE |

FIGURE 2.2 **The evolution of consumption**: Details the rise in consumer spending power and the impact on society and business

*Source*: Kjaer Global

lives on. For instance, Bernays was the first to tell the auto giants that automobiles should be sold as symbols of male sexuality; today we still see the echoes of that persuasion technique in the marketing of prestige cars.

Bernays was invited to promote the idea that ordinary people should buy shares, borrowing money from banks he was employed by. Millions followed his advice and suddenly he was known as the man who understood the mind of the crowd. Credit, and not savings, boosted corporate profits to new levels and access to easy finance meant that people spent money they didn't have. This resulted in a flourishing retail industry and helped create a stock market boom. Of course, there was a backlash when the Great Depression hit the US in September 1929, quickly spreading across the world. But even during Bernays' heyday and greatest power over the crowd, not everyone was seduced by this version of the American Dream. In 1927, an American journalist (identity unknown, unfortunately) made a rather profound statement about the changing role of the individual in relationship to the organization and the state, saying: "A change has come over our democracy, and it is called 'consumptionism', the American citizen's first importance to his country is now no longer that of citizen, but that of consumer."[10]

## Setting the measure of "enough"

This culture of consumerism has never gone unchallenged, but in recent decades there has been growing public debate about the ethics and moral compass of a society in which more equates to better. Endless opportunities for instant gratification have left many people feeling overwhelmed and ethically challenged, while others believe we have now reached a negative tipping point, where unsustainable lifestyle patterns are impacting and affecting us on all levels in society. In a 2011 LSE review of Diane Coyle's *The Economics of Enough: How to Run the Economy as if the Future Matters*, Joan Wilson addresses this:[11] "Many would argue that our relentless pursuit of higher economic growth, indicated through Gross Domestic Product (GDP) statistics, is at the heart of our current dire circumstances."

Coyle is right to question the metric on which we base our notions of prosperity, but it's interesting to note that others predicted a different course for our society. In the late 1920s – during the era when Bernays was busy setting the consumerist society agenda in America – economist John Maynard Keynes was working on the paper Economic Possibilities for our Grandchildren in England.[12] In this, he predicted that, by 2030, living standards in what he termed "progressive countries" would be between four and eight times higher and a 15-hour working week would be a reality because people would, as he put it "have enough to lead the good life." His prediction – one of the inspirations behind Figure 2.2 – looks highly unlikely today. Yet clearly, the question we need to answer before we can even consider the merits or likelihood of a Keynesian vision of the "good life" is: Who sets the measure of what is enough?

What Keynes didn't recognize is that an aspirational culture of higher living standards from an emerging middle class has meant we spent time chasing more and newer consumables. This – alongside the pressure to optimize everything, including our own productivity – has produced quite the opposite result to the one he predicted. Today, many people in Keynes' so-called "progressive countries" work harder than ever before, squeezing the maximum out of each working day while maximizing leisure time. Constant connectivity has resulted in "social jetlag,"[13] a phenomenon caused by the misalignment of our biological inner clock and our daily schedules, in which we attempt to do more than is physically possible.

Some even suggest that we have become sad slaves, not supreme masters, of the technology we invented to make our lives easier. To me, it appears self-evident that the quest for über-efficiency and hyper-consumption collides with our pursuit of "slow-living" and a balanced, sustainable society. Supply has created its own demand. Inexpensive clothes mean we buy more to wear, cheaper food means we eat more, discounted airfares mean we travel more, and mobile technology means we generate more conversations. In short, we have confused the "good life" with a life "full of goods" and consumables.

## The complexities of modern time

So the inevitable question must be: What does the new model for balanced consumption look like? Quality of life will always mean different things to different people, depending on geography and cultural context. But simply having enough space in the day to do the things that make us feel valued and productive members of society, such as quality time with friends and family and belonging to a community or network, are usually listed somewhere in people's definitions of quality of life.

Today's digital reality is one of constant flux and fluidity, influencing our lifestyle preferences. Technocrat behavior is now completely ingrained into our daily lives, with new technology developments demanding that we constantly nourish and feed our hyperconnected reality. This brings its own concerns, particularly about the consequences for the generations raised in front of a computer screen with a mobile phone in their hand. Some believe that this is making us feel less connected by weakening human-to-human bonds and interrupting traditional patterns of social behavior. An alternative perspective is more positive, since technological developments are part of a much broader set of changes and merely highlight our continued desire for connections and self-expression in a fast-changing and uncertain post-consumer society.

Although numerous media reports and studies have highlighted the absence of time in modern society, it also appears that we fail to recognize our freedom to chose when it comes to planning how to spend our lives. We have allowed work to blend with our personal environment completely, often mistaking action for progress by failing to set any kind of boundaries. Surrounded by supposedly timesaving devices, one of our most valuable commodities is, ironically, time, which has been monetarized and ascribed as a luxury. Although officially we work less than we did in the past, in reality we spend more time in work-related activities – be it commuting or being available on our mobile devices. There are certainly more interruptions in the day than there were 20 years ago; it is also true that proper time management and future planning often seem to be absent from our worldview.

What's wrong with the modern world, a 2013 *Guardian* article by American author Jonathan Franzen,[14] sheds light on this contemporary dilemma: "While we are busy tweeting, texting and spending, the world is drifting towards disaster." Franzen argues that, in our obsession with current media and technology, we are distracted from far more pressing concerns and have become unable to focus on anything but the present – meaning we forget the past and don't bother to imagine the future.

Even the idea of the future doesn't seem to make us dream anymore, perhaps because busy lifestyle patterns intensify our desire to seize the moment – any moment – and this is made worse by the proliferation of choice. Time used to be more rigidly framed through conventions about when to eat, work, shop, and so on; without this framework we have greater freedom, but there is also a price to pay for frenetic and fluid lifestyles. We can't turn back the clock, and nor am I advocating that; however, it is certainly worth contemplating alternative lifestyle suggestions. I am intrigued, for instance, by the New Economics Foundation's 2010 proposal for a 21-hour working week. The report, by Anna Coote, Jane Franklin, and Andrew Simms,[15] explained the reasoning thus:

> A "normal" working week of 21 hours could help to address a range of urgent, interlinked problems: overwork, unemployment, over-consumption, high carbon emissions, low well-being, entrenched inequalities, and the lack of time to live sustainably, to care for each other, and simply to enjoy life.

This proposal – which doesn't have mainstream political support as yet – would perhaps put our society back onto a footing where time can be enjoyed for its real worth rather than as a commodity. It would also take us a lot closer to the "good life" that Keynes confidently imagined back in 1930.

## Rethinking consumer mechanisms

So how did we get to this point? The idea of planned obsolescence as a business strategy was first employed in the 19th century and was revisited by Bernard London in *Ending the Depression Through Planned Obsolescence* in 1932. Engineers and designers were given new objectives to intentionally create products with a limited life cycle.[16] One example of this strategy is nylon stockings, once unbreakable and strong enough to tow a car and later manufactured to ladder so consumers would buy more. It is not difficult to see how this was a tempting strategy to keep the economy going; at that time, sustainability was not the issue it is today and the planet was perceived from the perspective of abundant, rather than finite, resources.

Perhaps it is no coincidence that the light bulb – the ultimate visual metaphor for ideas and innovative thinking – provides a textbook example of market manipulation through the obsolescence principle.[17] In the 1920s, the Phoebus cartel was founded by, among others, Philips,

General Electric, Tungsram, and Osram to create industry standards and control manufacturing and the sales of light bulbs. The 1000 Hour Life Committee was appointed to standardize the lifespan of light bulbs and members were fined if their bulbs' life expectancy exceeded the mark. Durability fell steadily and, by the 1940s, the target of 1,000 hours of life was reached – making this a classic case study of deliberately limiting the life and usefulness of products. The story has come full circle because Philips launched the EnduraLED bulb in 2012, said to last 30,000 hours or about 20 years. It is estimated it could save Americans $3.9 billion if every 60-watt incandescent was replaced, but with its high price tag at launch, this is unlikely to happen anytime soon.

In the 1950s, planned obsolescence became more mainstream, although not every manufacturer subscribed to the principle. Volkswagen created an advertising campaign to pitch its products as made to last, confidently stating: "We do not believe in planned obsolescence … good as our car is, we are constantly finding ways to make it better."[17] The cultural critic Vance Packard's *The Waste Makers,* published in 1960, argued that business was systematically engaged in making consumers wasteful, debt-ridden, permanently discontented individuals.[18] He wasn't the first to argue this, but he popularized the notion and his claims resonated with American consumers – the book remained on the bestseller list for months. While not a lot changed, the 1960s did see the evolution of early consumer movements, such as the Consumer Federation of America. The focus of consumer power at this stage was on protecting consumer interests in terms of guarantees and safety standards, but it was an early signal of the power consumers can leverage when they choose to join forces.[19]

However optimistic we may be about sustainable developments, Western society is still waste based. Indeed, our economies depend on replacements and upgrades rather than reuse – and to change this requires a mindset shift. But with a widespread desire to own something a little smarter, a little newer, a little faster, and a little sooner than necessary, change will not happen overnight. We have become primed into thinking that buying new is preferable to repairing or maintaining. And let's be honest, we are almost all guilty of this behavior. Jean Baudrillard's The System of Objects[20] explains it thus:

> The system of consumption constitutes an authentic language, a new culture, when pure and simple consumption is transformed into a means of individual and collective expression. Thus, a "new humanism" of consumption is opposed to the "nihilism" of consumption.

Certainly, Baudrillard has expressed one of the issues behind our society's obsession with acquiring ever better and newer products. Consumption remains a means of expressing ourselves, but we cannot ignore the indicators from our trend management work over the past 20 years, which reveal firm evidence that – for all industries – changing consumption patterns are set to shift the way organizations engage with their end users.

### SUMMARY: The evolution of consumption

- As the word "consumer" takes on increasingly negative connotations, there is an opportunity for authentic engagement by brands to help people make the right choices.
- For this to happen, companies must engage in honest, two-way dialogue with all stakeholders on how consumption evolves to ensure more people enjoy the good life.
- Hyper-consumption and pressure to optimize everything – including our time – means we live in the present and forget to plan for or imagine the future.
- A key debate centers on how we move from our waste-based economy to a more balanced approach, in which owning more is replaced by sustainable repair and recycle principles.

## Future consumer landscape and culture shifts

Consumption is an essential part of living, but it has also become a source of concern and sometimes guilt. We tend to polarize the debate by using the convenient argument that businesses are to blame – as they are the ones that should make it easier for us "to be good." This doesn't go deep enough into the dilemma because, as we all know, we transact freely and the businesses we turn to are made up of people – individuals who consume and make lifestyle choices just like us. In this regard, we are all in it together and must share joint responsibility for the consequences of production and consumption.

The negative impacts of our current consumption patterns have been well documented – increased pollution and reduced finite resources being the most persistent and persuasive arguments for altering our habits – yet most societies are still based on the attitude that consumption is what makes the world go around; it is our raison d'être. It would be naive to think this should or could be stopped in its tracks. But people have enormous power in their choices and this is becoming a tipping point for the many who have realized that conspicuous consumption is not the path to a fulfilled and empowered life. For that reason, businesses that haven't already considered the possibility that conscious capitalism is the way ahead must start exploring some soul-searching questions. Most crucially, they need to ask themselves: Is it possible for our business model to exist without planned obsolescence? And all of us – as producers and consumers of goods and services – need to consider how we can engage in a more truthful debate about current and future sustainability issues.

Moving further into the 21st century, it seems inevitable to me that the principles of efficiency and reuse enshrined in economic and ecological debates will radically alter production and consumption approaches. We are already seeing collaborative consumption[21] and the sharing economy in action through new disruptive business models such as Airbnb and Zipcar. Further evidence of grassroots change is in the proliferation of movements dedicated to changing the way we use and discard products. The Circular Economy 100 is an initiative by the Ellen MacArthur Foundation[22] to accelerate the transition into a Circular Economy. This global network platform brings together businesses, innovators, and regions to collaborate with academics and universities for collective problem solving on issues related to moving away from a "take, make, and dispose" economic model and into more sustainable reuse and remanufacture systems.

An early adopter on a larger scale in the UK is Marks & Spencer, whose Plan A strategy disrupted "retail as usual," with the tag line "doing the right thing." It was launched in January 2007 to address broader areas of corporate social responsibility. Encompassing everything from reducing energy and packaging to ensuring ethical sourcing and encouraging shoppers to donate unwanted clothing to the charity Oxfam, it has been a systematic project to make M&S "the world's most sustainable major retailer." The success of this commitment, now widely mentioned as leading in the field, is evidence that changing the game plan can work. Not only has M&S reduced waste by 28% and eradicated waste to landfill from its stores, offices, and warehouses, it is also delivering a profit from the savings back into the business – £135 million for reinvestment in 2012. Plan A has attracted plaudits across Europe and, notably, has assured M&S a place on the highly regarded Carbon Performance Leadership Index.[23] The company's approach is evidence today that major culture shifts are emerging at more than grassroots level to inform and reshape tomorrow's business landscape.

## The happiness factor

There is another persuasive business case for redrawing the map of consumption. Despite increasing affluence in the Western world over the past 50 years, we don't feel any happier,[24] suggesting that more affluence may not be associated with more contentment – in fact, quite the reverse. This is an issue Kjaer Global's trend-tracking activities reveal time and again; despite improved standards of living across the West, there is also increased dissatisfaction. Our lifestyle scanning shows that, without a true purpose and meaning to the consumption patterns they are encouraged to engage in, people feel increasingly empty inside. In *The Consumer Society: Myths and Structures*,[25] Jean Baudrillard says of the current state of play in our society that: "As much as it consumes anything, it consumes itself as consumer society, as idea."

Robert and Edward Skidelsky's book, *How Much is Enough? The Love of Money, and the Case for the Good Life*,[26] suggests that the "good life" can only be found if seven of our needs are met and turned into "basic goods" for everyone. Their list of seven comprises health, security, respect, personality, harmony, friendship, and leisure. Their take on the current situation is that if we didn't constantly strive to consume more, we wouldn't have to work so much and this would give us more time to enjoy life.

This gets us back to the concept and trend we call "enoughism." While the idea of setting limits on what we consume has been expressed by many commentators, I like to remind organizations of Mahatma Gandhi's words: "There's enough on this planet for everyone's needs but not for everyone's greed." So how can we mend the imbalance created by a culture of overconsumption? The answer is that we must evolve from production-focused economic models to more sustainable and democratic systems (from a ME approach to a WE approach, as briefly detailed in Chapter 1). With a far larger community of stakeholders influencing the direction of society and business, inevitably there needs to be shared responsibility and dialogue at all levels. When M&S established Plan A, it recognized that it needed not just to shift its internal culture, but also to engage its external stakeholders in the process of building a more sustainable future direction. This need for a transparent language is a constant refrain in the work we do with companies and I cannot emphasize its importance enough.

## Moving from short- to long-term thinking

Fundamental to the process of engaging with change is shifting from a pattern of short-term delivery to longer term rewards. This is quite a leap, as it requires us to overturn the traditional patterns we see all around us; most notably, we live within a political system based on delivering goals within a short time frame – typically the four- or five-year term of government office. This is one reason why I believe it is business and not governments that must take charge of the longer term agenda to deliver sustainability. In taking on this challenge, it has the monumental task of persuading shareholders and investors to look beyond quick returns on investment and consider the medium- to long-term future. I know of no magic bullet for that one, but perhaps the most persuasive argument is to remind all stakeholders that organizations – unlike politicians – need to plan for a future life that lasts well beyond four or five years.

As described in Chapter 1, the first step in trend management is to recognize that we should act on the future before it acts on us by becoming active change-makers. It is becoming increasingly apparent that external stakeholders – acting as individuals and as groups – are not going to allow the status quo to continue and every day we see signs of them challenging organizations to improve their policies and accountability. Social campaign groups, such as change.org, storyofstuff.org, and lovefoodhatewaste.com, are proliferating and their targets are far-reaching. Many of these organizations have already reached into the middle ground of consumers, talking in language people can relate to positively and with an aim to make everyone recognize their individual influence as a change-maker. The notion that anyone can start a global change campaign is certainly empowering for people; however, businesses that haven't prepared for this yet must recognize that their future depends on it. Radical openness, in which organizations open up and become transparent and collaborative, is on the horizon. Already, organizations have myriad new societal obligations and expectations to fulfill, so there is only one way to survive in an environment where accountability is the norm, not the bonus, and that is to demonstrate that you really do care by having a long-term business vision in place.

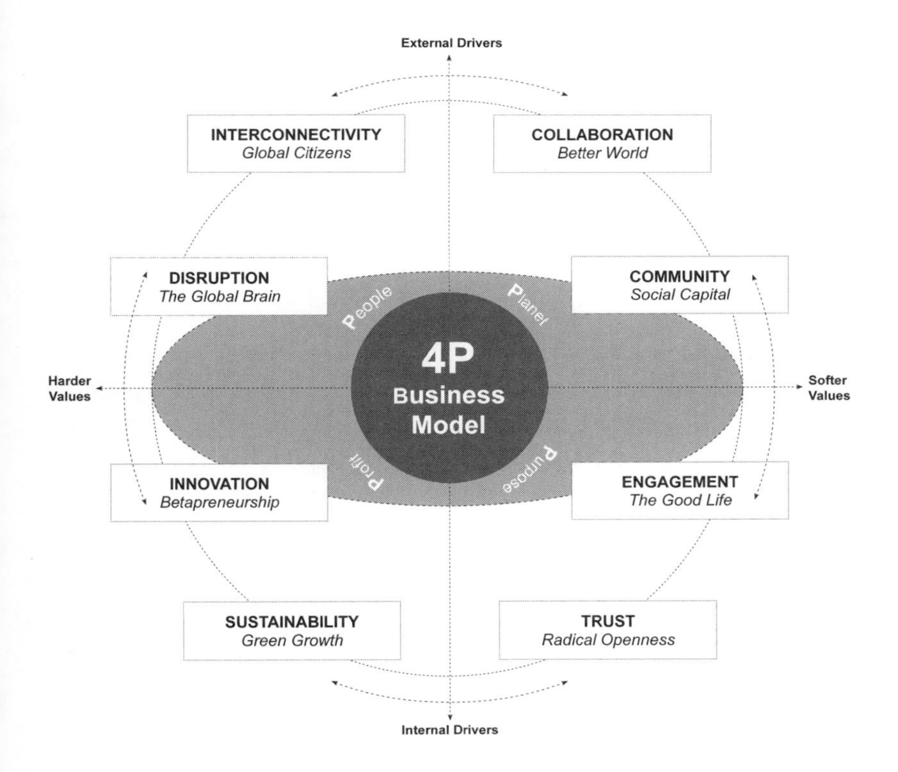


FIGURE 2.3 **The 4P business model**: A new economic ecosystem of people, planet, purpose, and then profit, illustrating the overriding key society trends and values driving tomorrow's business strategies
*Source*: Kjaer Global

## The capital P in leadership is purpose

Stakeholder capitalism and "doing good" are closely related to the "4P economic model" I proposed at *The Economist*'s Big Rethink 2012 conference.[27] Our four Ps are people, planet, purpose, and then profit – in that order. However we choose to assess tomorrow's business landscape, it is clear that "good life" aspirations are the social glue that links society, businesses, and people, and also the key to delivering purpose. Business success stories of the future will be those organizations that are agile enough to adapt to and balance the 4Ps so that they deliver sustainable, social, emotional, and economic value within one package. The 4P business model (Figure 2.3) illustrates how the relationship between these goals can work. Once we have a positive impact on people and the planet, with a purposeful ethos to match, we feed into our environment rather than feeding off it – and this is how we guarantee a place for our organization in the future.[28]

Organizations also have to recognize that in today's consumer-centric environment, decision making is becoming an increasingly heart-led process – a concept we initially termed Emotional Consumption in 2005, but evolved to Meaningful Consumption, choosing it as the overall theme for our 2006 Time to Think international conference in Copenhagen. People want to feel good about the products and services they buy into, and this is why emotional values have become the driver and rational values the passengers. This doesn't mean that "need" and "value for money" will become obsolete, rather that people are likely to prioritize personal sensations of wellbeing and meaning over material possessions – many already recognize that they don't want or need more "stuff" and focus instead on streamlining and downsizing possessions. When they do make a purchase, they consider the overall meaning and brand values and seek the opinions of trusted friends to ensure they make an informed choice (Figure 2.4). Almost two-thirds (63%) of people agreed that meaning and values are most likely drivers to influence their future purchase decisions.

## What do people want from organizations?

Undoubtedly, the biggest challenge for business will be to go from shareholder capitalism to stakeholder capitalism, in which the short-term maximization of profit and shareholder value is replaced by sustainable

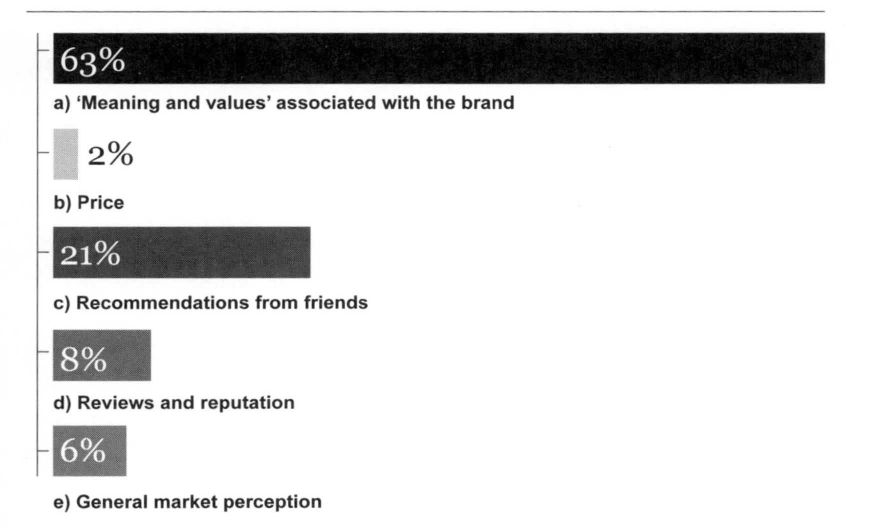


FIGURE 2.4 **Time to Think poll in 2006**: This survey forecast the growing importance of "meaning and values" in purchasing decisions
*Source*: Kjaer Global and Copenhagen Institute of Future Studies

growth and development (conscious capitalism). As Keynes' words remind us:[29] "The difficulty lies, not in the new ideas, but in escaping from the old ones." Social initiatives will play a key part in shifting us from "old" ideas, but to do this they must clearly demonstrate commitment around a common purpose to create social impact and genuine cultural change.

From a rational perspective, people want companies to provide goods and services through multichannel communication platforms, with evident transparency in systems such as payments, customer service, and feedback. But these objectives must be balanced with the softer emotional values of social capital and the "good life." These last two may sound like unquantifiable values, but they are precisely the challenges that Marks & Spencer is taking on through Plan A.

So how do we begin the process of balancing the rational and emotional needs of all stakeholders? On a large scale, Scandinavia is a good place to look for inspiration on how cultures can be designed to evolve around these key requirements. Already a global role model on happiness and prosperity, the Nordic region performs particularly well in innovation and social inclusion, and ranks top in economic competitiveness and happiness. A 2013 report in *The Economist*[30] said that: "the main lesson the Nordics can teach the world is not ideological, but practical." It's worth noting that, sitting in a geographically remote position and once considered among the poor cousins of Europe, Scandinavia didn't start with great natural advantages. But what it has developed through practical measures is robust internal ecosystems and long-term planning in order to develop self-sufficiency and sustainable prosperity. Some will question whether Scandinavian models can be replicated elsewhere, but undeniably they offer a vision of alternative approaches for managing economic growth to ensure inclusivity and more widespread wellbeing. Organizations need to foster a similarly far-sighted culture of thinking about and investing in an inclusive vision, and the journey into the future begins with getting the fundamental direction right through a structured process of trend management.

## SUMMARY: Future consumer landscape

- Principles of efficiency and reuse, enshrined within debates about a Circular Economy, look set to radically alter production and consumption models.
- Key to the business case for change is that increasing affluence in the West has not been a clear positive indicator for increased happiness but has led to dissatisfaction – inspiring the trend of "enoughism."
- Fundamental in the shift into the future is the move from short-term delivery to longer term rewards, in which we consider not just shareholders but stakeholders.
- Accountability is now the norm, so future business success requires a vision based around the 4Ps of people, planet, purpose, and then profit.
- Heart-led choices – or meaningful consumption – mean people are prioritizing engagement, so social initiatives are crucial to create impact and genuine change.
- From a rational perspective, people seek multichannel communication platforms with evident transparency in systems such as payments, service, and feedback.
- From an emotional perspective, people want social capital incorporated within the products and services they choose.

## The rise of meaningful consumption

There is growing evidence that people want to learn about themselves and other cultures by contributing to their local community and society at large. Fostered by network technologies, we are choosing to collaborate, co-design, and co-author to enhance our own personal ecosystem, environment, lifestyle preferences, and interests. This presents a real opportunity for companies to add value to their brand by facilitating and enabling people to reach their personal goals. This is an area emphasized by Havas Worldwide's *Prosumer Report: This Digital Life*,[31] published in 2013, which notes:

> We're seeing a push toward a "hybrid" way of living that combines the best of the old and new – keeping current conveniences while holding fast to those traditions and values that are in danger of disappearing.
>
> People are looking to replace hyperconsumption and artificiality with a way of living that offers more meaning and more intangible rewards – even as they wish to maintain the modern conveniences upon which they've grown reliant.

Situation-determined consumption means that people mostly make decisions based on how their choice or experience stimulates tangible and intangible needs. We are quite likely to consider the ecological or moral background of a company (the emotional pay-off), but our need for instant gratification remains a primary driver (the rational pay-off). When we consume, we want to feel a benefit; if we can feel good and know we are also doing good through our choices, that is a real selling point. This is why more far-sighted companies are already engaged in "betterness" programmes. Umair Haque, economist, author, and director of Havas Media Labs, has been a pioneer of betterness for businesses with his two books, *Betterness: Economics for Humans*[32] and *The New Capitalist Manifesto: Building a Disruptively Better Business.*[33]

Many trends point to the importance of intangible rewards and, increasingly, people's choices about what and how to consume are a tool to support an individual ecosystem they believe in. In effect, every time we buy into a product or service, we vote with our purse – supporting the ethical and social ecosystem of the brand. An important dimension of a democratic society is our right to vote – not only as citizens, but also as consumers. Companies need to be aware that more people are using this leverage to choose when to engage and when not to engage with brands. This means communication has to step up a gear, because at present the real disconnect between most companies and their stakeholders is communication – with an absolute absence of genuinely open dialogue. This can be addressed by not only listening and acting on customer feedback, but also considering the opportunities to take dialogue to the next level by assisting people in achieving happiness and meaning – and then inviting them to maybe trade with you again.

One of the great opportunities for 21st-century businesses is to collaborate and engage with people to help them do the right thing, as this builds social capital. Designing sustainable innovation and smart solutions is not straightforward – it takes time, effort, and authentic engagement to develop products and services that deliver on the 4Ps. However, this is an investment in the future, and if the only purpose of a product or service you introduce is P for profit, any short-term advantage must be considered alongside the potential damage to your long-term brand reputation.

In the 1980s, conceptual artist Barbara Kruger updated Descartes' original quote about the state of being human, "I think, therefore I am," to declare: "I shop therefore I am"[34] – a concise comment on our consumerist society. I believe that today it would be more accurate to say: "I feel, therefore I am" because it isn't just intellect that connects people to brands, it is also a "feeling"; in other words, our pursuit of meaning in the choices we make. Looking ahead in the 21st century, our Trend Management Toolkit suggests that our modus operandi will be much closer to: "I participate, therefore I am." This is because people are beginning to recognize the importance of their individual contributions to a stronger whole – by acting through their choices, they shape the society they want to see.

Organizations need to prepare for this evolution and start delivering meaningful experiences. By using the 4Ps as parameters, it is possible to instil a business mindset that focuses on purpose-driven leadership and close brand engagement, thereby becoming an organization that people want to be associated with. Businesses have to help people find meaning in the products and services they choose and the best way to start is to consider the values (the purpose) of your existence as a business.

Once you have determined that, you can then help people to feel they are making the right choices.

## Building real brand engagement

So how does an organization deliver meaning? The first step is to recognize that social networks are observing brands' every move. Online forums, comparison sites, and peer reviews are all engagement platforms for scrutinizing brands – they, and not you, are driving customer perceptions about your organization. This means that earning and renewing trust on a daily basis is essential and you can never bank on customer loyalty, but must prove you deserve it on an ongoing basis.

Our fundamental drive, the motivational engine that powers human existence, is the pursuit of meaning and I personally believe that this is a core driver of our 21st-century society. Given the myriad choices, design powered by emotion becomes essential, so organizations that offer authentic and engaging products and services will win out. People are now openly looking beyond the product to the experience; however, the degree to which this has always counted in our consumption decisions cannot be underestimated. In 1957, in *Motivation in Advertising: The Motives that Make People Buy,*[35] Pierre Martineau identified the undercurrents at work in our decisions, saying: "Any buying process is an interaction between the personality of the individual and the so-called 'personality' of the product itself." In other words, we want to like and identify with the brands we engage with in order to welcome them into our lives. It seems likely that, as we adapt to the changes the 21st century is bringing, we will increasingly set out to create our own personal lifestyle model that offers meaning, comfort, and, ultimately, satisfaction.

The Trend Atlas indicates that experiences that provide involvement, inspiration, and knowledge will be an essential component in the successful 21st-century brand fabric. This is about creating an environment where people are treated with respect as individuals – valued guests finding an enriching emotional connection with the retail experience and the products on offer. To put yourself in the customer's shoes – imagining the personality Martineau mentions as integral to the buying process – you need to create a comprehensive brand value universe. Your product or service must satisfy material, ethical, and value principles and, more than that, when customers ask themselves: "How does this product make me feel?" the answer is a positive. In short, your product or service must live up to all their expectations and make them feel good about themselves.

## Standing out from the crowd

Essentially, the more affluent we become, the higher our expectations are of the brands we buy into, and the current imbalance between our expectations and the experience we're getting is not positive news for many organizations. According to a global survey of 134,000 people, undertaken by Havas Media in 2013, only one in five brands are perceived as making a meaningful difference to our lives.[36] It is not that people are looking for utopia – they just want something better than the current flawed model, which brings us back to the importance of demonstrating a positive impact through brand actions and brand values.

In his 2010 book *People, Planet, Profit: How to Embrace Sustainability for Innovation and Business Growth*,[37] Peter Fisk suggests that organizations must ask themselves some fundamental questions underpinned by why they exist, why people will choose their brand, want to work for them, or invest in them. As Fisk suggests, conscious capitalism impacts significantly on what brands will be expected to deliver in the future. This means moving from green gestures and "me too" pledges to an ecosystem where you actively steward to improve every link in the chain in which you operate – from external producers and suppliers through to internal working conditions, packaging, global environmental impact, and social impact on your local communities.

Already we see intelligent and considered brand messaging from the likes of Puma, Patagonia, Marks & Spencer, Whole Foods Market, Unilever, and GE – to mention just a few. What these examples have in common is that they demonstrate a change-maker approach by seeking to elicit positive emotional responses and create a meaningful brand universe (Figure 2.5). Your loyal and potentially loyal stakeholders want to buy into and believe in your story and this means delivering a clear emphasis on the personal touch and the cultural context of your brand. Emotional engagement comes in many different forms, but it is important to recognize that people consider themselves in relationship to your brand and not the other way around. As Anaïs Nin pointed out in her 1961 work *Seduction of the Minotaur*: "We don't see things as they are, we see them as we are."

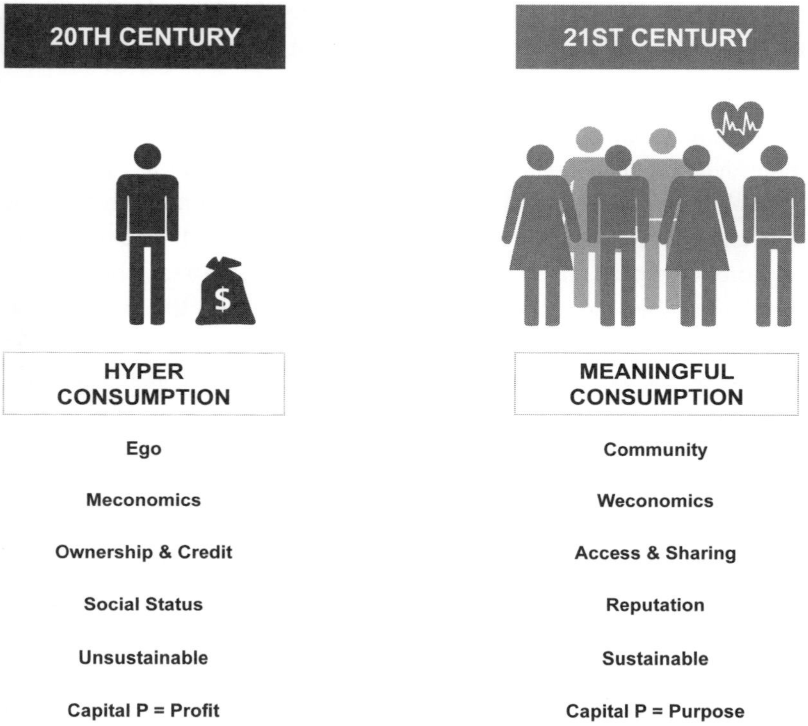

| 20TH CENTURY | 21ST CENTURY |
|---|---|
| HYPER CONSUMPTION | MEANINGFUL CONSUMPTION |
| Ego | Community |
| Meconomics | Weconomics |
| Ownership & Credit | Access & Sharing |
| Social Status | Reputation |
| Unsustainable | Sustainable |
| Capital P = Profit | Capital P = Purpose |

FIGURE 2.5 **Hyper consumption versus meaningful consumption**: Comparison between 20th- and 21st-century emotional motivators in consumption
*Source*: Kjaer Global

## Understanding the contradictions in society

Contradictions are everywhere and this is not an easy balancing act for organizations. Bombarded with mixed media messages and the constant pressure to reconcile urgent short-term decisions with a more considered, longer term strategy, business leaders are facing an information overload, where it can be hard to plot a steady or consistent path. None of us are dealing with a static market – people change constantly – so contrasts in the marketplace and consumer behaviors will always exist. This is why taking time out to think and reflect is such a vital step on the road to mindful business strategies that deliver consistent brand messages to the people we interact with. And, just as importantly, the process of taking time out to think means the strategies we develop will feel authentic to us – essential if we are going to deliver on what we set out to do.

Filtering and deciphering the interaction between people and product has become essential because it shows that people and brand relations are more dynamic than ever before. In the work we do, radical contrasts in people's behavior certainly exist – we see them all the time – so that it is perfectly possible for your audience to include people who want an ethically produced product, but at the same time expect it at a reasonable price. It has often been mentioned that cognitive dissonance – the gap between what we know and what we do – makes us unpredictable and irrational in the way we behave. The best way to make sense of these apparently disparate behaviors is by seeking to understand motivations, needs, and desires, and we do that by putting people at the center of the innovation process. This means looking at the world through their eyes to imagine what they experience in going about their daily lives and then creating scenarios that allow for these contrasts to coexist within the same value universe. The Trend Atlas research reveals these complex multilayers, and also exposes current and future opportunities for organizations – particularly in the areas of delivering positive guidance on making the right choices.

This brings us back to the importance of whole-brain thinking, which balances so-called left-brain pragmatism with right-brain creativity. We must also trust in the process – a subject I explore in depth in Chapter 3. As the Steve Jobs' Stanford speech reminds us, we need something more than just facts about the past to connect the dots of the future. Monitoring trends holistically requires us to tune into people's real needs and then create the meaningful interface they expect. People's priorities are changing – as is their level of trust for the institutions and organizations they deal with – so they need you to demonstrate your credentials. In effect, they are saying: "don't tell me, show me."

## Markers of change

An essential task on your journey into the future is to imagine possible scenarios, but another is to monitor and research from the bottom up, looking at ideas created by people in collaboration with organizations. People will always be at the center of everything we do because it is their patterns of living and consuming that set the future marketplace. For instance, when we start to think how people will live, behave, and spend their time in one or two decades, it opens up a whole raft of possibilities. Will we see more sharing communities or co-housing? Will people be commuting or working remotely? Will they demand green energy or cheap energy? Nobody can answer these questions with certainty, but we can look for "carriers" of emerging values – groups within society who look set to lend the next age its social profile, characteristics, and values.

Global Citizens are defined as culturally open individuals who are already changing our world. They are highly mobile, so businesses must be attuned to these change agents, who are setting new standards in virtually all areas of society. As digital natives, they are busy shaping new disruptive innovations and viable business models for tomorrow's society, organizations, and citizenship. In order to appeal to this group, new models must embrace intrinsic values, such as concern for others, community, and the environment, as well as delivering diversity that matches lifestyle requirements. This brings us back to the people, planet, purpose, and profit elements of the 4P model.

Global Citizens are set to be high-profile, demanding consumers, but they are not the only group that businesses need to engage with. In fact, defining and pinning down lifestyle groups is not the process it once was. Increasingly, we live in a polarized, "patchwork society" of people who share common lifestyles and value sets across conventional geographic borders. This means that businesses must move away from traditional rigid demographics into a more holistic understanding of people. Working with contrasting mindsets and social typologies to design solutions that can be adapted or used across the board is definitely challenging, but also provides a framework with a deeper understanding of future opportunities. Trend management sets the framework, enabling you to ask the right questions and posit the answers that enhance understanding of likely future changes taking place across all levels in society. Most of all, it helps us find the directions that will become meaningful and create commercially valuable solutions in the 21st century.

Through our work across multiple industries and geographic areas, what has really stood out over the years has been how trends develop and evolve. They don't come from nowhere, but often arise out of complex relationships and events in society. Decoding the relationship between socioeconomic drivers and the cultural contexts in society is seminal to forecasting, and perhaps the most inspiring part of the process, since it clearly shows how everything is interconnected. It is only when we begin to record, connect, and analyze the diverse drivers of today's cultural and economic landscape, as well as monitoring how they impact people's everyday lives, that we can start to translate research into synthesized future snapshots. These sound bites are the foundation of the Trend Atlas – a key instrument for deciphering future challenges as well as a tool to spot risks and opportunities for the purpose of converting them into viable business concepts. Using this multidimensional method has informed our work for so many years that we have a rich database to underpin and support our future visions. However, we are constantly reframing the boundaries as trends and drivers emerge or recede. It is this part of the process of trend management that provides a fascinating glimpse of how tactile the future actually is – also enabling us to consider, plan for, and help influence a changing world.

### SUMMARY: Meaningful consumption

- Individual contribution is becoming key, as people look to co-design, co-author, and collaborate – offering a real opportunity for companies to add value to their brand.
- Situation-determined consumption means people balance tangible and intangible benefits, with ecology and moral considerations coexisting with the need for instant gratification.
- Betterness programmes are the way forward in an era when social networks are driving customer perceptions about brands and people seek conscious capitalism.
- Trend management reconciles contradictory behaviors, as well as identifying "carriers" of future trends, such as the Global Citizens currently reshaping our world.
- Increasingly, we live in a "patchwork society," meaning businesses must move from rigid demographics to a more multidimensional understanding of people.
- Decoding the relationships between socioeconomics and cultural drivers is the fundamental basis of a Trend Atlas – a key tool for spotting risks and opportunities and new parameters influencing a changing world.

## CHAPTER 3

# Trend mapping: past, present and future

*It is evident that our society has, for far too long, banked on logical and linear thinking to the exclusion of creativity and intuition.*

As mathematician Henri Poincaré once pointed out: "It is by logic that we prove, but by intuition that we discover. To know how to criticize is good, to know how to create is better."[1] Emerging micro trends barely visible today might become critically important to tomorrow's worldview and great innovations. In this context, trend management informs our thoughts and ideas about what might happen by giving us a deeper awareness of the change drivers influencing society, our specific business sector, and people's behavior and values within this ecosystem. The most crucial activity in trend management is the ability to become attuned to shifts and changes, and the way to develop a full and rounded view of our environment is by exploring a broad and layered landscape of diverse drivers.

## From weak signals to macro trend

There is no magic to a trend – it is simply a steady uprising curve of an event or an influence that has the potential to become a powerful change-maker in society. Some trends reach a critical mass and become global (macro) drivers, while others remain a lesser influence operating on a smaller and more localized (micro) level.

Mapping trends is a way of looking at the key drivers of current and potential change, to observe how they have evolved over time, but at the same time consider how they are likely to develop in the future. To understand what impact trends are likely to have over time, we explore them in a 360-degree perspective of past and present influences. In other words, we need to engage in a multidimensional process and be willing to evolve and change the way we think about the world. To me, having been raised in Denmark and influenced by a society model firmly planted in the Nordic social-democratic tradition, using brainstorming, workshops, scenario planning, voting, and polling are approaches that come naturally as methods to develop insights into the dynamics of trends and their relationships on a micro and macro level. Such approaches are not always practiced routinely within organizations, but they are among the most widely used methods for tapping into the future, as outlined in *The Handbook of Technology Foresight: Concepts and Practice.*[2] The tendency towards using qualitative methods – looking for deeper insights and taking a lateral approach – has become commonplace when organizations consider the future, according to the European Foresight Monitoring Network (Figure 3.1).

## Rethinking human decision making

Human decision making is a complex cognitive process, since numerous elements drive behaviors and therefore have the potential to profoundly influence choices, on a conscious and unconscious level. In this and the next chapter, we explore various factors influencing decision-making processes among groups and individuals. But first let's consider how data, media, and visual stimulation have been used over time to influence decision-making processes by tapping into people's emotional and rational landscape.

As discussed in Chapter 2, the 20th century saw the emergence of tools invented purely to influence the minds of the crowd and sway decision processes, and during America's mass industrialization Edward Bernays masterminded strategies that achieved this goal. One of his most dramatic was an early attempt to capitalize on the growing undercurrent of female

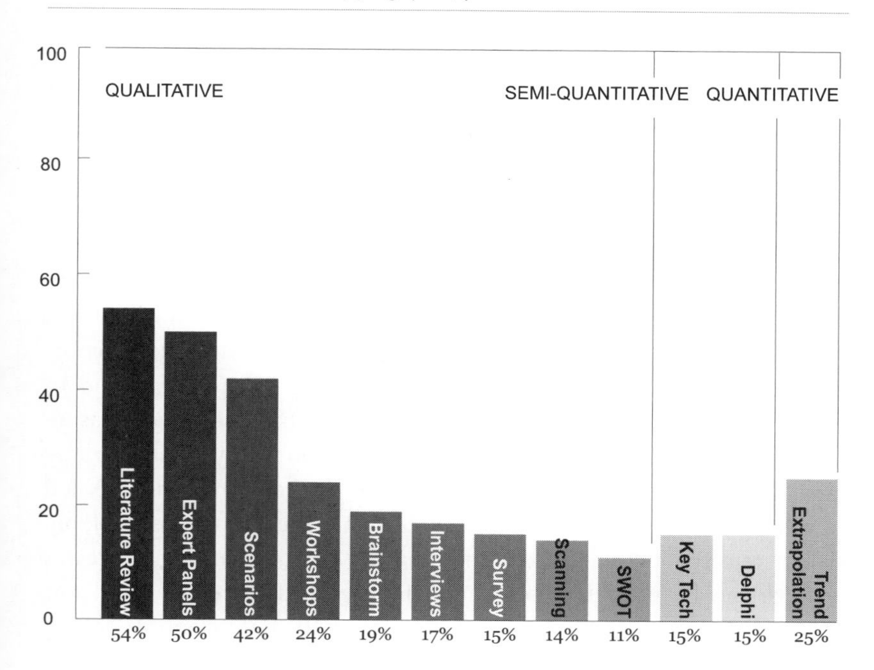


FIGURE 3.1 **Most used forecasting methods**: The most common approaches are all qualitative
*Source*: Kjaer Global. *Data*: European Foresight Monitoring Network

empowerment after World War I. He sought to persuade women to smoke in public, flouting social taboos of the time, by hiring a group to march during New York's 1929 Easter Parade and light their "torches of freedom" to signify equality with men.[3] While we may question the morality of this tactic on many levels from today's perspective, it was a striking early example of how existing societal undercurrents may be spotted and successfully harnessed by organizations.

Later, Bernays' techniques were applied on a grand scale in published advertising and on billboards, as well as in newspaper editorial sections, to reinforce notions of the consumer society. In an interview in 1967, German philosopher Herbert Marcuse[4] said of Bernays' ideas: "This is a childish application of psychoanalysis which does not take at all into consideration the very real political systematic waste of resources of technology and of the productive process." Marcuse further argued that if people were reduced to expressing their feelings and identities through mass-produced objects, it would result in what he described as "one-dimensional man" – a conformist and repressed hostage to a consumer society.

Psychoanalysis was not the only force in play during the era of systemized influencing of public opinion and buying patterns. In response to the Great Depression of the 1930s, President Roosevelt launched the New Deal to promote the idea that democracy and capitalism went hand in hand. Roosevelt invited social scientist George Gallup (the inventor of opinion polls) to help him explain his policies to the public and to take their opinions into account. Gallup claimed that one could measure and predict the opinions and behavior of the public by asking strictly factual questions that avoided manipulation of their emotions. This led to weekly public opinion polling to report what the nation was thinking. Gallup rejected Bernays' view that human decision-making processes were driven by unconscious emotional forces and therefore could not be trusted if too many choices were presented. His scientific polling established the counterargument that people are rational, they do make good decisions, and democracy is furthered if people are given a voice to influence how the country is run.

## Cognitive traps and the focusing illusion

A key assumption that took root throughout the 20th century – especially in economics – was that people base their decisions on rational judgment, thinking about long-term goals. However, recent research shows that emotions and intuition play a major role in our decision making, leading us to focus on short-term personal goals. This is why many organizations are now employing behavioral economics to understand how conflicts of interest may bias our decisions and perception of experiences. This discipline – which merges social, cognitive, and emotional research with traditional economic theories – considers the cognitive dissonance between what we know and what we do; in other words, it explores reasons why we may be hardwired to think in the short term rather than carefully consider the consequences of our behavior in the long term. These findings have had profound implications for government, public policy, and future economic models – and our own self-awareness.

One of the world's foremost authorities on behavioral economics is Daniel Kahneman, and in a 2010 TED lecture entitled The Riddle of Experience vs. Memory,[5] he shared his insights on cognitive traps. Using the happiness trend as an example – one that almost everyone talks about nowadays – he points out that the term "happiness" is applied to too many things, adding complexity to the meaning of the word. In addition, the dissonance between experience and memory results in two very different concepts in the notion of happiness. Kahneman suggests that being happy *in* your life and being happy *about* your life are two quite different things. Another theory related to this is what he terms the "focusing illusion." In this, our present experiences and memory present quite different pictures. In effect, we act as twins when it comes to making sense of our lives – one living and knowing life in the present and the other maintaining previous memories and our life story. Kahneman's theories would suggest that predicting people's behavior and reactions is a far less cut-and-dried business than simply deciding whether to follow the Bernays or Gallup model (unconscious or conscious). In fact, the truth is that we cannot be classified into simple behavioral patterns because humans are complex, as reflected in the diversity of our lifestyle choices and the memories and experiences we choose to collect.

If we step back for a moment and apply this complexity perspective to trends in general, it reinforces why it is so important to take a broader view and look at potential trend topics from numerous angles. When we consider current thinking about human behavior, it soon becomes clear we live in a landscape of many interesting yet contradictory theories about what motivates people and their actions. But we must also recognize that we are influenced by structural and behavioral drivers, even in talking about trends, as Adam Gopnik pointed out in a September 2013 article for *The New Yorker* entitled Mindless: The New Neuro-Skeptics: [6] "The neurological turn has become what the 'cultural' turn was a few decades ago: the all-purpose non-explanation explanation of everything." Gopnik adds that: "Psychology is an imperfect science, but it is a science." I would extend his conclusion to all scientific theorizing about predicting future behavior; this is why it is important to be aware of the scientific and social contexts in which we operate, as well as our emotional and spiritual landscape – it enables us to take the broadest possible perspective when planning ahead and imagining our collective future.

## New thinking on left versus right brain

While we do not process information using the left or right brain in isolation, this shorthand metaphor helps us to assess the kind of reference framework and style people utilize in their everyday lives. As part of our polling and information-gathering activities at conferences and workshops, we often ask participants how they process information, and their responses tend to run true to what we would have predicted. When a group of risk analysts were asked: Are you a left- or a right-brain thinker? at a recent risk conference, almost a third said that they see themselves as factual, logical, and pragmatic left-brain thinkers, while just over half described themselves as mostly using both left- and right-brain thinking. Designers tend to describe themselves as favoring the right, while 61% of internal communication officers say they balance the pragmatic left brain with the intuitive strengths of the right brain (Figure 3.2). Perhaps no surprises here, but it should be added that we are witnessing interesting trends at work. To return to risk analysts, just a decade ago a vast majority would see themselves as left-brain thinkers – suggesting that more emotional value parameters are now invited into the equation, even among traditionally rational-skewed professions. This is certainly a key personality question every reader of this book should ask themselves: Do you favor intuitive decision making by tapping into an emotional/feeling open-thinking mode (right brain), or do you prefer factual rational/logic closed-thinking (left brain), or would you describe yourself as utilizing both (whole brain), depending on situation and context?

In his book *Thinking, Fast and Slow,*[7] Daniel Kahneman redefined the left-/right-brain idea as fast System 1 and slow System 2 and suggested that we use the entire brain to process information. His assertion that neurons are firing up and connecting across the whole brain is what we see in neuroimaging or fMRI (functional magnetic resonance imaging). This is a method used to map neural brain activity and it shows how cerebral blood flow and neuronal activation are coupled when an area of the brain is in use. This network system of processors helps us catalogue and make sense of the information we receive or experience: visual, sound, smell, taste, and so on. The truth is that, while we know a lot more about our minds than we did just a decade ago, this is still a fast-developing science with many more mysteries about brain function and power yet to be discovered.

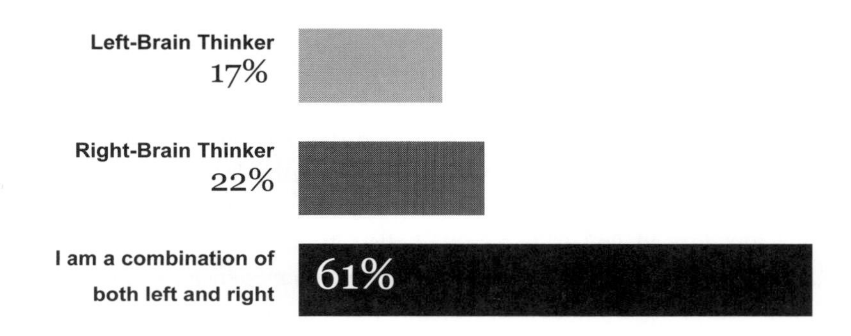


FIGURE 3.2 **Left- versus right-brain thinking**: Most communications professionals perceive themselves to be whole-brain thinkers
*Source*: Kjaer Global

In 2011, psychiatrist and writer Iain McGilchrist gave a lecture at the Royal Society of Arts on his book *The Master and his Emissary: The Divided Brain and the Making of the Western World*.[8] He remarked that the division of the brain is something neuroscientists don't like to talk about anymore, as the first split-brain operations in the 1960s and 70s led to a sort of popularization of the left-/right-brain dichotomy, which has since proved to be entirely false. McGilchrist noted: "both are profoundly involved in everything we do."

According to McGilchrist, we have one operating system, but the nature of left- and right-brain mechanisms is that they offer two versions of reality. One is knowledge mediated by the left hemisphere, functioning within a closed system and driven by perfection and detail. The other, facilitated by the right hemisphere, delivers broader understanding and sense making. Hence, we need to rely on certain brain skills to navigate the world and others to make sense of it. As McGilchrist puts it: "There is a paradoxical relationship between knowledge of the parts and wisdom about the whole." He further suggests that we have turned into a left-brain society that honors the artificial over the "real thing" and uses Einstein's quote to sum up his own vision that: "The intuitive mind is a sacred gift and the rational mind is a faithful servant." McGilchrist goes on to describe the

21st century thus: "We have created a society that honors the servant but has forgotten the gift."

## The interconnected brain

While countless personality tests, team-building exercises, and self-motivation books have been built around the popular belief that people are either left-brain or right-brain dominant, myth-busting scientific studies are revealing a far more intriguing mystery at work in our cognitive makeup. Indeed, in 2013, neuroscientists from the University of Utah analyzed more than 1,000 brains and found no evidence to suggest that people favor either side of their brain – instead, their analysis suggested a "whole brain network." In the article Researchers Debunk Myth of "Right-brain" and "Left-brain" Personality Traits,[9] Jared Nielson commented:

> Everyone should understand the personality types associated with the terminology "left-brained" and "right-brained" and how they relate to him or her personally; however, we just don't see patterns where the whole left-brain network is more connected or the whole right-brain network is more connected in some people. It may be that personality types have nothing to do with one hemisphere being more active, stronger, or more connected.

If, as this study suggests, our style is based on our personality type, not our wiring, then this gives us far more control over the way we approach decision making and planning for the future. As argued in Chapter 1, critical thinking about the future is a skill that can be developed when we choose to engage whole-brain approaches. Ongoing work looks set to revolutionize how we understand our cognitive skills. Already, scientists are moving traditional 20th-century notions forward. In The New Science of Mind,[10] an article for *The New York Times*, Eric Kandel described the developing field of the science of mind. This combines cognitive psychology and neuroscience to solve one of the last great mysteries – how we think, feel, and experience the world. Kandel argued that: "This new science of mind is based on the principle that our mind and our brain are inseparable." The data below is selected from the OECD (Organisation for Economic Co-operation and Development) in *Understanding the Brain: Towards a New Learning Science*[11] from 2002 about popular "neuromyths." Each statement confirms some essential insights about learning and, if we have to summarize the findings in one line, the clear message is: we never stop learning.

> NEUROMYTHS
>
> **Learning has a limited time frame:** Brain plasticity is not limited to early years and learning therefore happens continuously and causes the brain to form new connections at any time in life.
>
> **Enriched environments in early age enhance learning capacity:** More research is needed, but since our brain is plastic throughout life, it seems logical to deduce that neural connections can be established at any time.
>
> **There are visual, auditory, and haptic styles of learning:** While these learning styles (eyes, ears, touch) may be important for initial perception, we use all our brain for processing information and then moving to intellectual understanding.
>
> **We only use 10% of our brains:** Activity and mapping techniques have shown that all brain regions are active, even when we are asleep.
>
> **Multilinguals "lose" capacity in one language and can't transfer knowledge:** Many people operate in multiple languages, and there is no scientific basis for the idea that our brains can't accommodate or switch between languages.
>
> **We favor the left or right brain:** Scientific evidence shows that, while there are functional asymmetries, our brain hemispheres do not work in isolation but operate together in every cognitive task.

## Popular brain myths debunked

Before we explore the multidimensional thinking style to future forecasting in more depth, it is worth considering a few common misconceptions about our brain function. Many myths imply that our brains are static and can only support a certain amount of information. But evidence points to a counter reality – neuroplasticity – meaning that our brains adjust to accommodate new information, whatever age we are. A University College London study undertook MRI scans on London black cab drivers before, during and after they had undertaken "The Knowledge" (the mentally demanding two- to four-year process of memorizing routes around London). This showed that the posterior hippocampus – connected to memory – had grown during the process of memorizing 25,000 streets and 20,000 landmarks.[12] This is just one finding, but it's an area where many beliefs are being overturned.

### SUMMARY: Rethinking human decision making

- Trend management helps us understand the influences on people's behavior, values, and decisions. It is important to consider global (macro) and local (micro) trends, since both influence business and society.
- The most crucial factor in trend management is to become attuned to shifts by exploring a broad landscape of cultural and societal influences, considering past and present influences in a multidimensional perspective.
- We now know that emotions and intuition play a major role in decision making, leading to a focus on short-term goals, and this suggests far more complexity in human behavior and choices than previously thought.
- While rational (left-brain) thinking helps us understand the details about a challenge, we need intuitive and visionary (right-brain) approaches to consider the bigger picture in a whole-brain manner.
- Recent theories suggest we have more control over decision making and future planning than previously thought. In addition, proof of the neuroplasticity of our brains suggests that learning is a flexible and lifelong process.

## 21st-century sense making

While we may once have imagined that more data would make it easier to understand behavior and reach informed decisions, precisely the reverse is true. Indeed, the present data deluge has also become an Achilles heel – where the vast quantity of facts and figures from multiple reliable sources that should be our strength is making it harder to sift and process information and narrow it down to a meaningful choice. To understand today's multilayered world, we need to move trend mapping forward so that it complements the surge of available source data while also taking into account our changing environment and the lifestyle patterns influencing people in the 21st century.

So how do we become smart decision makers in a landscape of unprecedented change and complexity? Our memory and logical reasoning can serve as indicators of how we arrived at our current point, but relying on only one kind of thinking for important decisions will not give us a visionary outlook for the future. As already discussed, rigid and evidence-based left-brain methods provide us with a specific view of past patterns, but to go deeper about why these patterns happened in the first place, we must include right-brain visionary synthesis to make robust narratives about the future.

Navigating these opposite forces means we need a simplified abstract version of reality, one practiced within multidimensional forecasting. Although we know that the brain is not divided but indeed interconnected to form a complex intricate network, the left-/right-brain dichotomy serves as a useful metaphor in a world in which overreliance on facts and rationality is still the norm to the exclusion of intuitive and creative thinking. To contemplate trends in a fully rounded way, we need both spheres working in tandem as this forms the foundation for a whole-brain approach.

In Chapter 1, we classified the different forecasting methodologies as scientific, social, emotional, and wild card, but it is useful to consider them within the multidimensional thinking model (Figure 3.3).

## Left-brain personalities: scientific and social

Analytical left-brain methods are essential for breaking the whole into components in order to closely examine the details. These may be expressed in terms of pure benefits, numbers, and bottom lines. In this realm, scientific forecasters use logic and factual data that is largely evidence based; while social forecasters employ similar factual approaches, their data is mainly drawn from interactive evidence gathering.

## Right-brain personalities: emotional and wild card

Intuitive right-brain methods synthesize fragments by weaving them into a whole in order to assess the bigger picture. This approach looks beyond data to consider quality of life in a broader sense. Emotional forecasters use creative, imaginary, and intuitive-based approaches; wild card forecasters (the spiritual dimension) focus on using imaginary scenarios to deliver expertise-based high impact but low probability predictions.

Returning to McGilchrist's theories on how our brain processes information, it is helpful to use the analogy between the brain and a computer. Like a computer, our brain has one operating system but we employ different "software" for different work tasks. Brain processing (Table 3.1) provides an illustration of the key differences between left-/right-brain processing styles.

TABLE 3.1 **Brain processing**: Illustration of left- versus right-brain processing styles

| Left brain | Right brain |
|---|---|
| General | Individual |
| Fixed | Changing |
| Static | Evolving |
| Isolated | Interconnected |
| Denotative | Implicit |
| Lifeless | Incarnate |
| Decontextualized | Living being in the living world |
| Analyzes | Synthesizes |
| Breaking the whole into components | Weaving components into a whole |
| Focusing on details | Seeing the bigger picture |
| Benefits and bottom lines | Quality of life in a broader sense |

*Source:* Kjaer Global

## Multidimensional thinking explained

In essence, multidimensional thinking is a timesaving method that draws on components of all the main forecasting styles in order to access a whole array of information in our society. Hard values presented by scientific facts and current social forces are tempered with a more instinctive understanding of the inspirations that drive us to a particular worldview or consumer behavior. The multidimensional method operates as a platform for trend management and a trend-filtering system to facilitate building a Trend Atlas. In this process, we also utilize psychographics to delve into people's emotional landscape and explore behaviors. It is an approach where we study and measure the attitudes, values, lifestyles, and opinions of our audience for the purpose of creating personas or social typologies.

As we have already explored, most forecasting methodologies overlap and draw from each other, as illustrated in Figure 1.1, Key forecasting methodologies. Rational analysis of a trend development provides a sound basis for speculation and prediction, but in today's unpredictable climate, multiple ingredients are needed to define the crucial building blocks for

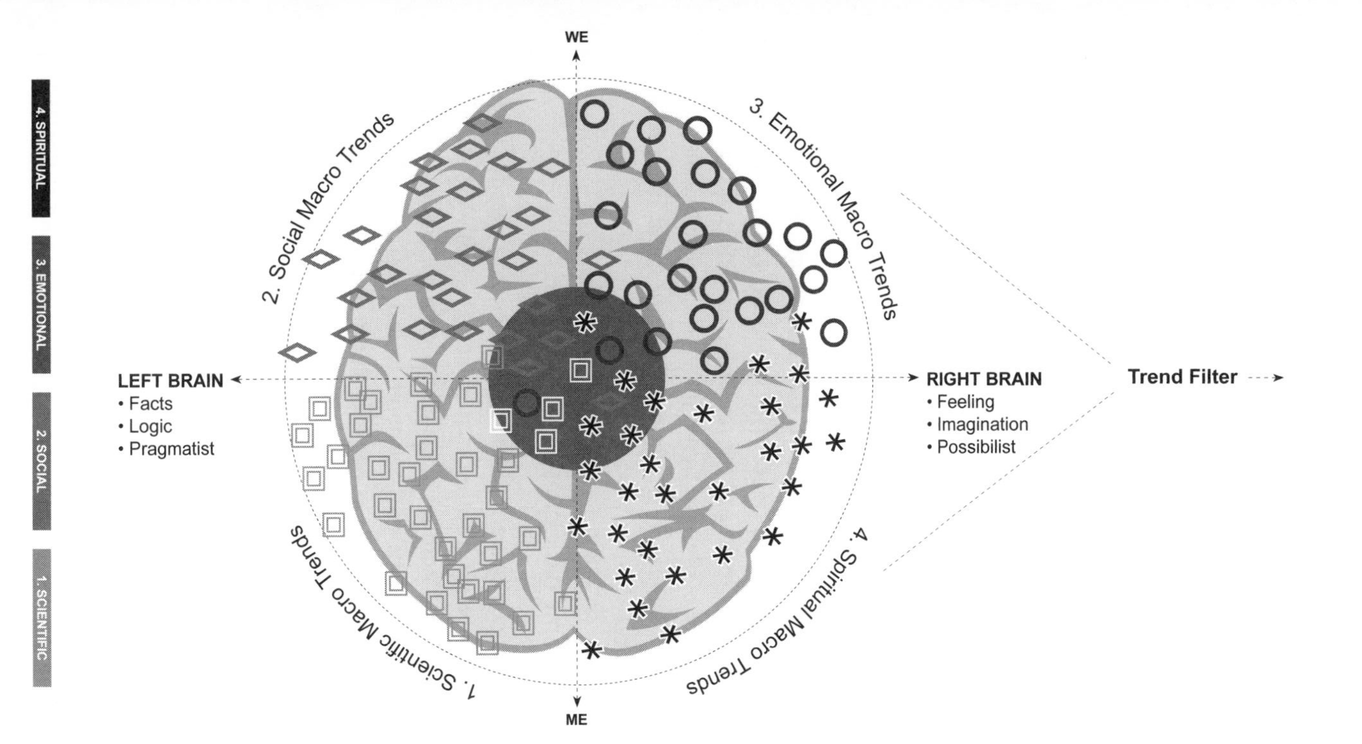


FIGURE 3.3 **Multidimensional thinking model**: To understand how society, businesses and people are interconnected, our model considers multiple dimensions

*Source*: Kjaer Global

creating meaningful future scenarios and road maps. Reasoning methods must therefore be robust enough to observe trends from a multilayered standpoint, balancing facts with informed intuition.

The 4D quadrant and multidimensional Trend Index (Figure 3.4) acts as a lens and a filter for plotting and synthesizing data captured during the research phase. The quadrant has left and right hemispheres, divided into four dimensions, classified as scientific, social, emotional, and spiritual (explained in depth in Chapter 4). When framing trends for future scenarios, we usually choose between 8 and 12 key macro trends from our generic Trend Atlas (see Figure 3.5 below) – typically 2–3 for each dimension – to ensure a balanced representation of the trends. The supporting micro trend drivers are then selected – 2 for each macro trend – typically they are informed by local insights, topics or specific influences to provide a wider and more meaningful context. For more comprehensive future scenarios or deeper exploration of a particular topic, it is possible to extend the number of micro trends. This is dependent on the number of macro trends being explored and the research parameters of the organization. When the number of micro trends increases, it will invariably present a deeper and more detailed insight into the specific topic; however, it will also add complexity.

## Rethink and reframe basic questions

Emphasis on productivity, performance, efficiency, and speed is no longer helping us move towards a future that makes sense. Always striving for more and better – and without actually mapping the value and purpose of our endeavors – means that we tend to lack clearly outlined goals. Consequently, there is an urgent need for businesses to rethink and redefine the map of success by addressing profound questions about growth, prosperity, sense making, and quality of life now and in the future. As we explored in Chapter 2, accountability is the norm and a 4P approach – with "purpose" defined clearly in strategy – is the starting point for this re-evaluation process. To actively influence the future, it is essential to allocate time for quiet reflection, especially in a world where value is more commonly placed on everything tangible and quantifiable, to the exclusion of more visionary approaches. Not only does time to think help us move away from a culture focused purely on output and reward, but it

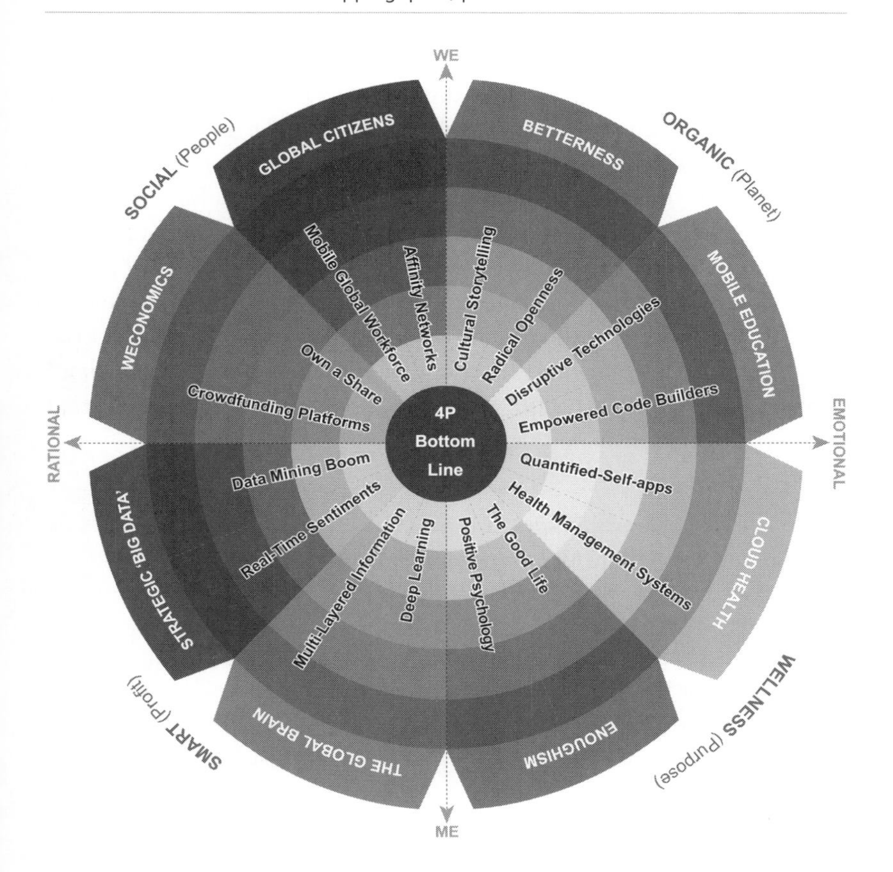


FIGURE 3.4 **The 4D quadrant and multidimensional Trend Index**: This chart, populated with macro and micro trends, illustrates how to connect core drivers selected from the Trend Atlas
*Source*: Kjaer Global

also enables us to find alternative – and possibly better – answers to the pressing challenges we face, as businesses and as a society.

One clear avenue for discussion and ideation is in the meeting ground between science and arts/humanities; and it is interesting to note that many leading scholarly organizations, including the Wellcome Trust and Dana Foundation (UK), Max Planck Institute (Germany), and Smithsonian and MIT (US), have stepped tentatively into this open space by holding multidisciplinary public debates and events where academics from different specialist areas exchange ideas and theories. This runs counter to the approach that has, over the course of the past two centuries, increasingly separated the research/academic world into tangible (science) and less tangible (arts/humanities) silos. Such forums are profoundly stimulating and may be the missing link in our society that will help to foster the intuition and creativity excluded from conventional and more insular research centers, which – by the nature of the process and standards of academic rigor – have a tendency to create independent avenues of theory/knowledge. In the process, such meeting grounds may also take on the issue of deeper purpose that must be considered in any global debates about people's future lifestyle, health, wealth, and aspirations – and the role that business plays in satisfying these needs.

### SUMMARY: 21st-century sense making

- The quantity of data we are exposed to makes it harder to sift information and narrow it down, meaning we must adapt trend mapping to take into account not only source material, but also changes in environment and lifestyles.
- While rational (left-brain) thinking helps us understand details about a challenge, we need intuitive and visionary (right-brain) approaches to consider the future of society and people.
- The multidimensional method – considering scientific, social, emotional, and spiritual influences – enables us to build a platform for trend management and filtering, using a multilayered approach to build a robust set of scenarios and a "road map" to guide future strategy.
- It is vital to rethink our future, reframing basic questions about our end goals, considering quality of life and the 4Ps of people, planet, purpose, and profit.
- The meeting ground between science and arts/humanities, currently being explored, is opening up global debate on the issues surrounding people's future lifestyles and the role of business in meeting those needs.

## From 2D to 4D thinking explained

To make the quantum leap from a two-dimensional (2D) to a four-dimensional (4D) vision, we must step outside our organizational box and consider a much broader perspective of elements at play in society and culture. The problem with viewing the world in just 2D is that we fail to look at the bigger picture – hence our organizational reference frame becomes a relatively narrow window of knowledge.

First let's consider why 2D thinking is not enough. A 2D view focuses on the road ahead and considers current and future influences through the prism of the specialism – usually a specific industry or sector. It's effective at providing a multitude of information about market conditions, known competitors, and so on. However – and here's the problem – it is perfectly possible to scan your market scrupulously with scientific exactitude, adding in a few traditional demographic insights, but chances are you still miss crucial emergent or disruptive elements coming from the area of the horizon outside your frame of reference. Alternatively, you may spot a subtle new pattern in the data but dismiss it as a "weak signal" because your tools and research parameters don't enable you to understand it in any greater depth.

Most notably, this approach fails because it does not include people in real-life situations or factor in their individual needs and desires. However, this is not a new discourse, but an ongoing one, because ultimately it is not just about new technology and available data, it is about corporate culture and purpose. Engaging with people and acquiring the kind of data that is useful in a 21st-century context requires a different approach, as we will discover in Chapter 6. The multidimensional thinking model was developed in order to address this shortfall in traditional forward planning, also matching clients' need for "living and breathing" future narratives that can be used as a tool in the innovation and change management processes. On a practical level, it operates by tapping into the collective memory and individual perspectives of an interdisciplinary team. By bridging the disparities between internal departments, it encourages dialogue and empowers people with different areas of expertise to speak and understand the same language and then move forward together in order to formulate a purposeful and informed strategy.

## The Global Brain and our collective memory

In *Smarter Than You Think: How Technology is Changing Our Minds for the Better*, Clive Thompson sets out reflections and observations on how the Internet is a tool that can shape and influence people in a positive way. He argues that this Global Brain (the Internet of Things) that connects us all across conventional geographic borders is also shaping new forms of human cognition and making us smarter by interconnecting everyone and everything:

> Understanding how to use new tools for thought requires not just a critical eye, but curiosity and experimentation. ... A tool's most transformative uses generally take us by surprise. How should you respond when you get powerful new tools for finding answers? Think of harder questions.[13]

Today's reality is that our technological tools are now inextricably linked to our minds, or at least working in tandem with them, profoundly changing how we learn, remember things and, as Thompson points out, how we "act upon that knowledge emotionally, intellectually, and politically."[13] This can be viewed as a promising development rather than a negative consequence; it provides the opportunity for individuals to collect knowledge from day-to-day experiences, gain a broader worldview, and make better decisions. And for the group, there are similarly encouraging signs. Transactive memory was a theory developed by Daniel Wegner in the 1980s as a response to earlier hypotheses about the "group mind."[14] He suggested that the whole becomes greater than the sum of its parts because, via group or transactive systems, we tap into other people's knowledge and expertise. It is a system that operates within families, social groups, and even large organizations and enables a memory system hugely more powerful, potentially, than any one individual's knowledge. Both these theories suggest that another of today's realities is that we are able to process far more information than was once thought possible – we have the opportunity to think and act smarter.

## The essential future building blocks

Tapping into the future is not a linear thinking process, but rather a multilayered exploration that requires us to connect multiple points of reference. This method helps individuals and groups ask the right questions about the future of their organization – using collective memory and also drawing on individual expertise and life experiences – to make better decisions. The combination of individual and collective memory or knowledge can be used to create an organization's future road map by plotting the trends most relevant to the immediate environment. An example of

FIGURE 3.5 **Generic Trend Atlas 2030+**: This tool maps the trends in the scientific, social, emotional, and spiritual dimensions of society (see Figure 4.2 for full size on pp. 100–1)

*Source*: Kjaer Global

*Images*: New Models: "Mapping Design for Circularity," Credit: The Great Recovery Team at the RSA; Green Growth: Credit: Samsø Energy Academy how this works to collate relevant information is shown on the generic Trend Atlas (Figure 3.5), and the processes involved in mapping trends are explored in depth in Chapter 4. This initial framework for mapping macro trends guides the process of plotting future directions, with macro trend knowledge increased and underpinned when additional micro trends relevant to your locality, your market or specialism are added.

The 4D framework serves as your checklist – ensuring that you get a balanced outlook to assist with formulating strategy. The generic macro trends are essential building blocks for mapping the future as they form an overview to kick-start any project with a medium- to long-term time frame. By selecting the key trends affecting your organization – distributing them evenly in the 4D model – you create your GPS to the future using a selection of viable narratives as the starting point. This visual overview inspires and enables the analysis required to move from 2D to 4D, see the bigger picture, and consider challenges and opportunities. It is essentially a pattern-making tool to assist with informed decisions in the present about the future.

## Using maps in trend forecasting

Throughout history, maps have been instrumental as a common reference tool to help explain complex concepts and ideas, share knowledge about the known and unknown world, and teach people how to navigate from one point to another. Maps contain a wide range of complex information, rendered in an open visual form, and may also serve to rewire our perceptions, showing connections and combinations we hadn't spotted before and enabling us to discover new solutions and fresh ways around old problems.

There have been many inspiring examples of maps and map making throughout history, but in more recent times, cartographers and designers have begun using technological tools to create amazing visual representations of information to explain and explore complexity. This process is also called "data visualization," often represented as network diagrams or infographics to compress and simplify data. The big trend of the moment is to illustrate complex information on geographical maps, by overlaying real-time dynamic data to reveal interconnectedness and relationships.

In effect, maps are a visual representation of diverse data and have the potential to change the way we think about the world. By viewing two points on a chart, we may suddenly recognize that navigating from point

A to point B can be done in various ways. As a method to ask questions about the future, maps can influence our sense of identity and place – even change our purpose and direction. Their use in trend forecasting is to provide instantly accessible meaning and clarity. The best-written 200-page report is of no earthly use if you need to absorb a lot of information fast or convey it in summarized form to someone else. In effect, a map – the trailer not the epic – sets the scene and provides the context that enables you to understand the implicit meaning. A good map, which is always underpinned by sound exploration and research from a variety of perspectives, has the potential to serve as an orientation tool, providing us with a quick overview about the world and our environment.

### SUMMARY: From 2D to 4D

- Moving from 2D to 4D (multidimensional) approaches requires us to step outside our organizational box and consider a wider window of knowledge. Scanning only your own market (2D) excludes people and potentially disruptive market and society events.
- The Global Brain (Internet of Things) is enabling us to work and think smarter as individuals and groups using transactive memory, in which we combine forces to gain knowledge and learn from each other. In effect, this is broadening our worldview.
- The Trend Atlas, like all good maps, helps us to navigate complex information. It is essentially a pattern-making tool to assist with informed decision making in the present about the future.

## The future is not set in stone

A crucial element of trend management is to recognize that there is not one but several possible scenarios for the future. First and foremost, whether we operate as a government, a business or an individual, our choices influence and shape the future. While this may appear challenging at first, once we consider and map out likely outcomes of the most relevant future scenarios – considering risks and opportunities – the process becomes liberating. If we haven't already done so, it is at this point that we recognize the value of stepping away from business as usual to start shaping purpose-driven future strategies and directions – making new and better choices. In other words, we progress to the point where we recognize our own individual and organizational impact as change-makers.

Representations of possible directions are often termed "future scenarios." As briefly described in Chapter 1, the first seeds of scenario planning were planted in the early 1940s by the American analyst Herman Kahn in contemplating likely outcomes of various nuclear strategies. In an interview with *The New York Times* in 1982, Kahn himself expressed it thus: "We draw scenarios and try to cope with history before it happens."[15] This method of narrating the future was later evolved within corporate settings, and has since become widely used by business and governments. These future narratives are all about experimenting in order to create imaginary maps of the future, to develop and support choices of strategy.

## Developing future narratives and scenarios

Trend management scenario development is a powerful tool for creating dynamic content that opens minds by exploring events that might affect the overall environment (external) and our ecosystem (internal). The mechanics of this process are agile and flexible; we extrapolate data and insights, mix current ideas with new thinking, and then scale them up and down. This might be as simple as replaying a future scenario with one component altered – such as imagining what happens if the price of a resource rises or falls – or as complex as we care to make it by adding in a variety of elements. For instance, we could test our future scenario if we experience a talent shortage and an innovation that disrupts our market. Specific aspects of the scenario may be emphasized and others completely left out in order to develop alternative visions for our sector or industry.

The wide range of possibilities unlocked by trend mapping is what helps to ensure persuasive scenarios that meet essential criteria; however, it's important to also separate out what we think might happen and what we hope will happen. These are termed "realistic" and "idealistic" scenarios. Static or dynamic scenarios are realistic, while the optimistic or radical visions are more idealistic. Ideally, these scenarios counterbalance each other and several are needed to gain a multidimensional insight into the future. In practice, the narratives often overlap and evolve, ending up as a balanced combination of all four; or alternatively, we choose to work with two very different scenarios.

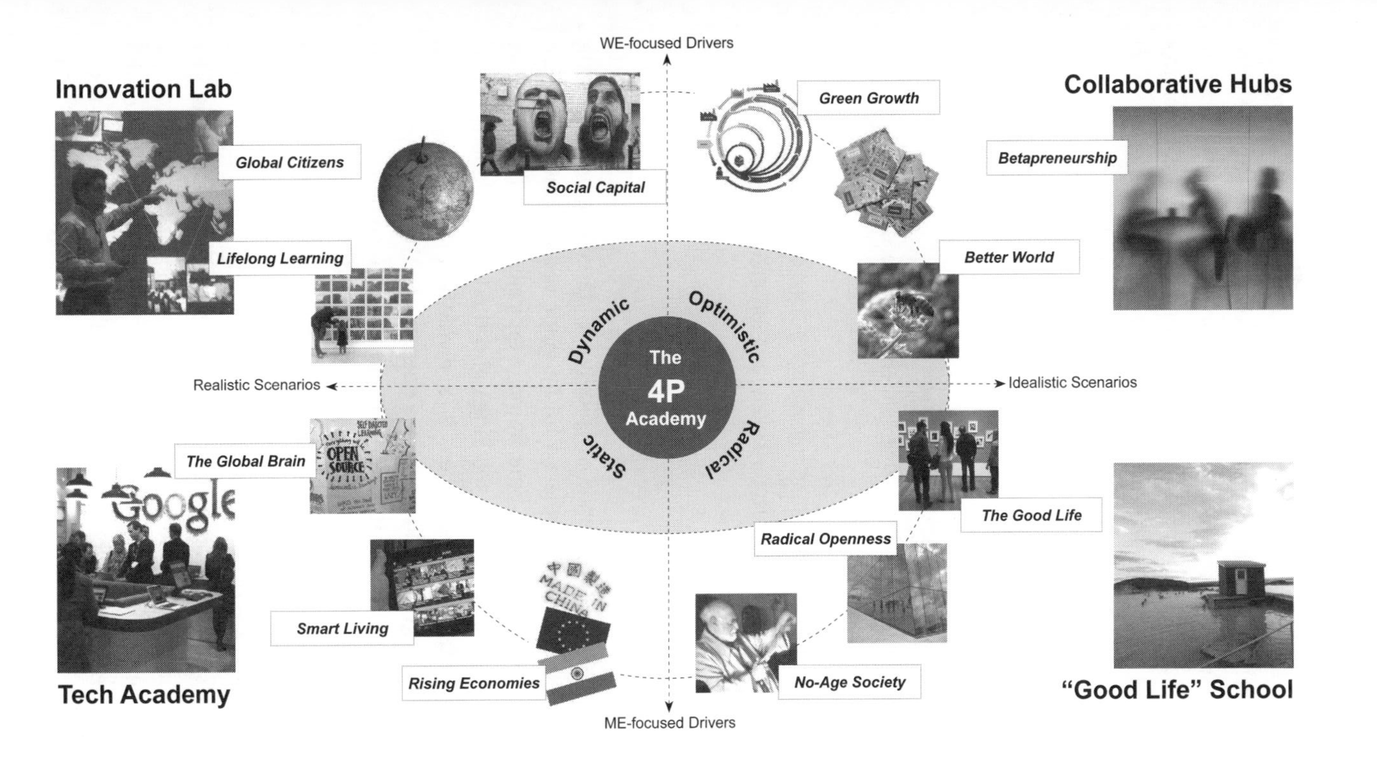


FIGURE 3.6 **Future scenarios example**: The scenarios overlap, but eventually end up as one balanced combination of all four or two alternative outcomes

*Source*: Kjaer Global

*Image*: Green Growth: "Mapping Design for Circularity," Credit: The Great Recovery Team at the RSA

The Future scenarios example (Figure 3.6) illustrates what a scenario map may develop into – naturally, individual sectors have their specific parameters or indicators related to the environment in which they operate. The most crucial part of scenario building is for organizations to connect the "dots" or trends most relevant to them. This trend-mapping process is a collaborative one, where the group or team discuss and agree on what trends to select from the Trend Atlas. We will explore this a little deeper in Chapter 4.

## REALISTIC VS IDEALISTIC SCENARIOS

Realistic scenarios:

**Static:** Most likely, conservative and/or predicable.

**Dynamic:** Significant and known emergent/disruptive trends or events.

Idealistic scenarios:

**Optimistic:** Strategically different to competitors or significant market opportunity.

**Radical:** Overturning fundamental beliefs and assumptions with persuasive alternative.

## The process of shortlisting trends

This process of shortlisting trends is typically facilitated in a brainstorming session or a trend workshop where a future guide will facilitate discussion and agreement over a limited time frame – typically three to four hours divided into exploration (open thinking) and then consolidation (closed thinking). In instances where there is a divide within a group trying to select key trends, it is valuable to go deeper into the source as it may provide clues to specific threats or opportunities surrounding the disputed trend/s that are already evident in an organization's overall strategy or direction. Methods and frameworks used in trend management should always be tailored to meet the specific objectives and end goals of an organization, as we outline in Chapter 4 by exploring how to perform a Trend SWOT (strengths, weaknesses, opportunities, threats) analysis. The role of the facilitator or future guide is to direct/mediate the process to highlight key areas – this can be done ahead of the brainstorming session or workshop, usually by asking a set of future-directed questions sent out to the group digitally. Alternatively, the organization or group are asked to submit questions they would like answered in the future workshop session.

## Communicating the future

Several trends should be combined to imagine possible future implications for the organization or particular sector, which is usually done by creating a tailor-made Trend Atlas or map of experiences. It is important to note that when we present learning and insights to stimulate future strategies, reports and presentations built on facts and figures tend to resonate solely with rational thinking individuals; for this reason, adding visuals and visionary diagrams is crucial because it makes future scenario material engaging and relevant to a much wider audience. Visuals that illustrate and support content under discussion are not only more inclusive, but can improve learning by up to 89%.[16] In *On Photography,*[17] author and political activist Susan Sontag reminds us that visuals can expand our processing power:

> Photographs alter and enlarge our notions of what is worth looking at and what we have a right to observe. They are a grammar and, even more importantly, an ethics of seeing. Finally, the most grandiose result of the photographic enterprise is to give us the sense that we can hold the whole world in our heads – as an anthology of images.

Perhaps the most important take-away when choosing your preferred method for documenting and communicating the future is that it should be used as a diverse information system that inspires the analytic and creative members of your team, or, in our shorthand, the left- and right-brain thinkers.

### SUMMARY: The future is not set in stone

- In future planning, there is not one but several scenarios. Our choices shape tomorrow and we have the scope to change outcomes by making better decisions.
- In scenario building, it's important to distinguish between realistic (static or dynamic) and idealistic (optimistic or radical) projections. We consider both to ensure a balanced overview of potential futures.
- Broadcasting the future begins with a collaborative process of trend mapping, in which the organization connects the "dots" or trends most relevant to them via a brainstorming session or workshop.
- Several trends are combined to imagine possible future scenarios using a tailor-made Trend Atlas. Where a group can't agree on what trends to shortlist, it's

important to explore the disagreement as it may highlight existing threats or opportunities to an organization.

- Methods and frameworks must always be aligned to the organization's end goals and trends should be combined in order to imagine a variety of possible future scenarios.

## Signposts to the road ahead

As discussed earlier in this chapter, one of the key challenges of academic research is that disciplines have become highly siloed, while societal challenges are multidimensional. The impact of this segregation of knowledge into neatly independent units, so unlike the messy, overlapping, and contradictory world we inhabit, is currently the subject of a profound rethink. The London School of Economics (LSE) 2011 *Maximizing the Impacts of your Research: A Handbook for Social Scientists*[18] touches on one of the most pertinent challenges in sharing knowledge:

> While academic departments, labs, and research groups produce a great deal of explicit knowledge, it is their collective "tacit knowledge," which is the most difficult to communicate to external audiences, that tends to have the most impact.

The LSE also sheds light on the challenge of relying solely on linear thinking in business; the problem being that value measured is always prioritized over value captured: "In short, metrics or indicators can tell us about many aspects of potential occasions of influence, but not what the outcome of this influence was."[18] By default, these metrics have become the main driver of business and they do give us easy-to-assess results. But clearly they don't tell the whole story and as the LSE suggests, there is a need to combine linear and lateral thinking in the 21st century.

With the adoption of arts/sciences bridging by many renowned universities and institutions, it seems evident that the future will bring more interdisciplinary collaboration as people group together as problem solvers – a scenario where rigid thinking is replaced by agile modular thinking. This agility is even more striking in the rise of MOOCs (Massive Open Online Course), with Coursera just one of many organizations offering a whole new way of learning and collaborating online. The same approach must be adopted by organizations that want to think smarter in the 21st century. Using a cross-collaborative tactic to bridge disciplines will be a crucial means to promote integrated thinking and engage in more efficient and inspiring ways of working and developing fresh ideas.

## New thinking models

If you want to tap into and learn more about the true needs of tomorrow's people, traditional demographics are no longer pertinent. A holistic approach is essential, which means implementing thinking from a wide variety of research and design practices to develop long-term planning and fertile scenario building. Drawing on quantitative and qualitative thinking forms a whole-brain synthesis that acts as a framework for future innovation.

The capability of people-centric and social-centered research continues to expand to solve global challenges and it delivers high hopes for solutions that will define the future of innovation. Among corporations and business schools, there is an emerging understanding of the potential of visionary lateral thinking in research, but this will require a much wider curriculum in academic institutions and a clear social focus. Arguably, we need to start the process in early years education to enable the next generation of scholars and academics to think less rigidly. Currently, these principles are being applied among early adopters within communities of innovation and by academic pioneers, such as the Singularity University, founded by Ray Kurzweil and Peter Diamandis. Its stated mission is: "to educate, inspire and empower leaders to apply exponential technologies to address humanity's grand challenges."[19]

## Seeing the future in a new light

As we are now well into the second decade of the 21st century, it is time that we looked to fresh thinking and new models as a way to drive more sustainable economic growth and advance human welfare globally. Organizations and businesses already working with scenario planning know that seeing the future in a new light is the only way to discover viable solutions and recognize untapped opportunities.

Coming back to the landscape of global challenges we are all facing, it is interesting to compare them with some of the key findings of the OECD International Futures Programme. This suggests that, by 2030, daily life will seem radically different for large parts of the world's population when compared to the 1990s, while technological change will have transformed what it means to be human. "Smart mobile everything" is already a reality, one in which our lives are streamed directly to and from the Global Brain, adding simplicity, autonomy, and possibilities to supply better health, education, and opportunities for personal fulfilment to many more people.

Major socioeconomic and geopolitical transitions are happening; the challenge of our times is to bring the unfolding reality of these forces into line with what people really need and want. One key to surviving as a business in the 21st century is to ensure that strategy and innovation models are shaped to explore the risks and opportunities likely to arise during the transitions into this new reality. For businesses, the imperative is both real and pressing, and critically needed for rebuilding trust, which is in historically short supply for large corporations. Businesses have a balancing act to achieve: to uphold their social and ethical responsibilities in inspiring and educating their customers to live more sustainably while still delivering a profit.

## SUMMARY: Signposts ahead

- Sharing knowledge is crucial to the journey of planning future strategy – bridging the gap between different departments and disciplines in order to share knowledge and experience is crucial to develop agile and modular thinking.
- With increasing evidence that new structural models are set to develop in the 21st century, business strategies should focus on the risks and opportunities that are likely to arise during the transition period.
- There is an emerging understanding of the potential of visionary lateral thinking in research to move us towards disruptive innovation with a social focus.
- Ultimately, we must move business into line with what people really need and want. This means a strategic balancing act: upholding social and ethical responsibilities, inspiring customers while still making a profit.

## Chapter 4

# Your essential trend toolkit

*Developing a 360-degree outlook is the only way to understand the bigger picture and actively shape the future as a society, a business or a citizen.*

One of the most empowering things about working within the Trend Management Toolkit is that it does not act as a prescription for how your organization must think and act. Instead, it scans a four-dimensional landscape of scientific, social, emotional, and spiritual trend influences in order to inspire a visual representation of how the future might evolve and manifest. As discussed in Chapter 3, the system is an interdisciplinary method that draws on elements of all the key forecasting approaches, also allowing room to think the unthinkable – the wild card predictions – enabling companies to develop a bespoke outline or disruptive idea for their future.

The multidimensional forecasting approach is a structure to encourage the development of a range of scenarios, utilizing a method that recognizes and explores the ways in which markets and the people of tomorrow are likely to develop. The main objective is to make informed decisions about the needs and wants of tomorrow's people. Current trends will always evolve and take on new meanings, depending on shifts in society, political development, and environmental challenges, while technology and socioeconomic progress alter lifestyle patterns. To address and explore these shifts, we have developed a Lifestyle Navigator (see Figure 6.11) that enables the cross-referencing of macro and micro trends, consumer behavior, needs, wants, motivators, and values to expose the opportunities that these insights present when considered in a holistic context.

## The bigger picture

Following the digital revolution, things no longer change over a generation or a decade, but from year to year, even month to month, creating new arenas for disruptive ideas and innovation to emerge. Seismic events since the start of this millennium are contributing to a new world order, not only challenging our value landscape, but also our notions of what constitutes economic and social progress. Inevitably, this leads us to reconsider the future and our place within it, and hence imagine possible alternative futures.

The accelerating pace of change is fueled by digital technology developments – particularly in the way we communicate and learn about the world – and this progress calls for much more intelligent but also intuitive systems to process information. Despite a growing data deluge, we still know very little about people's emerging needs and wants, and keeping pace with analysis and pattern spotting to extract useful data means pinning down what is, effectively, a moving target. This makes it essential to develop the right tools for converting data insights into meaningful analysis and relevant strategies that are attuned to 21st-century fast thinking. Currently, we are witnessing how global dynamic forces influence our behaviors – a reality where "mechanical and rigid" are being replaced by "organic and agile" – profoundly influencing people's belief systems and value universe. This is a point that will be expanded in Chapter 5, in which we explore the key structural society drivers and trends that are influencing all levels of society.

### Society is driven by macro and micro trends

Defining the parameters in which societies are organized and operate is an ongoing discourse and a multifaceted debate. Therefore, let's start by clarifying how we define the world, and our role within it, for the purposes of trend management. Structural drivers are the foundation of societies, driven by a set of change agents, which are made up by macro and micro trends. The macro trends are distinct because they operate on a larger scale, while micro trends generally happen in a narrower context in terms of time frame, geography or field. In trend management, we typically observe the micro level to forecast how the macro level might evolve over time – meaning that a global macro trend might have evolved from within a micro context that eventually reached a critical tipping point. Big (macro) changes are invariably made up of lots of smaller or micro shifts operating within a local ecosystem. Basically, structure is static while change is dynamic – the drivers that make the world go round.

Initially, a futurist explores the big picture, also called "environmental scanning," within a framework of key structural drivers, such as the political, economic, societal, technological, environmental, and legislative landscape, to understand how these affect the given sector or organization. This initial analysis, termed PESTEL, is described in greater detail later in this chapter. However, the structural drivers beyond PESTEL must be balanced by the value drivers of community, culture, and ethics. Trends are interconnected and can't necessarily be confined or classified within one distinct category. Nonetheless, this useful framework sets the scene for exploring and selecting the most relevant macro trends, which will shape and inform long-term developments.

To be beneficial, a future trend management system must inform and inspire an organization across departments – from board level to R&D, design, marketing, and communication – underpinning innovative strategies to build the foundation for a common future vision where everyone takes ownership. This means that, further through the process, it becomes possible to project impacts in your particular market and then use these insights to envisage scenarios with practical relevance to your organization.

## Tools for mindful reflection

The big issue for businesses is that, all too often, they are using yesterday's tools to solve today's and tomorrow's challenges. Developing a 360-degree outlook means greater insight into potential challenges and opportunities for society, businesses, and the individual, but this perspective does not emerge if we only analyze qualifiable data, we need to look beyond. Today, most organizations are starting to recognize that adopting a strictly rationalist approach to a nonlinear set of circumstances inevitably inhibits the ideation of innovative strategies and solutions. Our specific contemporary challenges – notably empowered and increasingly demanding consumers, a plethora of new and disruptive business models, and rapidly changing lifestyle patterns – mean we need tools to open up dialogue and minds to the new world order. And once we accept that uncertainty is the norm, we can move forward to create engaging and believable future narratives.

The Trend Atlas not only epitomizes multidimensional thinking, it also acts as a structuring tool in trend scanning, selection, reflection, and analysis. It combines the efforts of the individual and the team to produce profound in-depth research and data collection. Input from experts and opinion polls in a variety of fields may be used to add depth by giving additional visionary or practical insights to the initial research. Qualitative – as distinct from quantitative – market research is valuable as a means of capturing multiple nuances and then delivering a detailed insight into individual perspectives to inform the core analysis. This process of overlaying and comparing data cannot be rushed, requiring mindful reflection, observation, and application, but when done correctly it opens up a world of future possibilities that can set a business on course for innovation and sound future strategies.

## SUMMARY: Bigger picture

- Accelerating change – fueled by digital technology developments – fuels disruptive ideas and innovation, inspiring us to consider alternative futures.
- Analysis and pattern spotting, in the midst of a data deluge, requires a multidimensional (360-degree) approach, drawing on elements of all key forecasting approaches and also allowing room to "think the unthinkable."
- By considering scientific, social, emotional, and spiritual shifts, it is possible to gain an informed and balanced view of tomorrow's people.
- It is vital to distinguish between micro and macro trends, by recognizing how trends may manifest in different socioeconomic and cultural contexts.
- To be useful, a trend management strategy must inform and inspire an organization at every level, ensuring everyone takes ownership over the future vision and can consider practical implications.
- The Trend Atlas not only epitomizes multidimensional thinking, it also acts as a structuring tool for trend scanning, selection, and analysis.

## Creating your Trend Atlas

Trend management is, first and foremost, an inspirational and practical thinking tool for making the future more comprehensible. The process and underlying research method is typically conducted in four phases: scanning and collection, analysis, synthesis, and, finally, communication.

### Scanning and collection

It is vital to isolate key societal movements to understand the undercurrents, motivations, and likely developments ahead – in a socioeconomic and cultural context – so that you can pair and then connect them. The research and collection phase provides a plethora of information and knowledge, ranging from literature review, insights, and facts and figures to interviews and case studies. Detecting key trends from among the astronomical number of undercurrents and influences requires patience and an open mind, particularly as there are multiple layers of insights to consider in a big data reality where new network landscapes are emerging all the time. The endeavor of identifying trends pays off, as this process gives insights into the direction of key society drivers and the potential outcomes of macro trends. Crucially, the process also provides pointers about tomorrow's lifestyle choices and, when considered in conjunction with trends and drivers, generally reveals new ideas and possible solutions to longstanding problems.

The generic Trend Atlas enables you to navigate the initial data complexity by presenting findings in a global and local context. At first, the trends can appear intangible, but when you overlay findings you typically start to see weak signals that eventually emerge as clear patterns. The trends you selected for a particular location, sector or industry will eventually support and validate each other once you have clustered them, by logically forming the foundation for scenario building. The multidimensional platform classifies trends on a macro and micro level, enabling interpretation of their importance in each of the dimensions. As stated before, it is important to remember during this process that the dimensions do not exist in strict isolation, but are interconnected and sometimes overlap.

## Soft analysis and core analysis

The analysis is divided into two parts: soft analysis and core analysis. The initial soft analysis narrows down the many potential influences to extract a core group of macro trends and micro trends, paired and coupled within the multidimensional framework to highlight the most significant elements that will play a fundamental role in shaping the future in a specific setting.

## Future building blocks

Each macro trend is typically influenced, or underpinned, by a number of core micro trends and the constellation is supported by insights, facts and figures, general impact, and case studies. The macro trends are fed into the Trend Atlas framework – typically they form an initial map to provide the bigger overview. From the soft analysis, the process moves on to a core analysis to consider the trends in a more people-centric perspective, where trends are crystallized into easy-to-navigate future building blocks.

### 4D METHODOLOGY

**4D (four-dimensional) scanning:** Used to frame investigation and conduct research. This broad "net" of interconnected insights considers the cause and effect of trends on society, businesses, and people in a macro and micro context.

**The generic Trend Atlas (Figure 4.2):** Provides the grammar and context of forecasting and serves to inspire ideas. The core principles of the Trend Atlas are explained later in this chapter.

**A historic overview of society, organizations, and people:** Typically, a 20-50-100-year time frame is useful for setting the scene and considering the context in which your organization works. The evolution of consumption diagram (Figure 2.1) provides a simple chronological meta-perspective of our world and provides a useful starting point.

**A sector time frame and overview:** Highlight key forces that have shaped your specific industry or organization in a more general perspective. Do not underestimate the importance of considering your unique heritage – where you have come from and how you have arrived at this point – since this provides a solid foundation for projecting your organization's future.

**Collaborative process:** This work is mostly done by a team of interdisciplinary practitioners and futurists, preferably in close collaboration with the organization leadership team to ensure the Trend Atlas is aligned with internal processes and R&D activity.

## Synthesizing potential futures

For the final synthesis, we develop a universe of personas projecting social typologies, consumer situations, and values in relation to the particular industry or sector we are exploring. In order to work towards a tangible action plan and conclusion, it is important to map out potential futures, behavior patterns, and needs in this phase. Initially, we create a visual structure known as a "consumer Mindset Map" (explored in Chapter 6) that works as a template for building viable future scenarios.

## Communicating core findings

The final phase of any successful trend management activity is communicating the core findings across the organization. The communication phase provides best value if findings are shared across the organization so that core stakeholders can understand and utilize the findings. Typically, this is achieved through group presentations, workshops, and Q&A activities, inviting everyone involved to take ownership and make change happen.

## Understanding the foundation

The Trend Atlas offers a system of interconnected trends that provides a dynamic overview of the future. New trends emerge or shift to different dimensions on the map progressively, as society views evolve and alter. Before we can evaluate and monitor trends over time, we must first decide which indicators we want in our Trend Atlas framework – an exercise of placing significant macro trends in each of the four dimensions.

### Quantifiable factors

The first two dimensions, scientific and social, relate to the factual influences in society by tapping into left-brain values. These are quantifiable and usually widely publicized factors that all of us are familiar with from day-to-day life, through awareness of current affairs and knowledge of our own market.

### Qualitative factors

The other two dimensions, emotional and spiritual, defined by right-brain values, capture the qualitative and more intangible forces at play in society. This is where we observe trends from a people-centric perspective of consumer behavior and attitudes, considering culture and lifestyle patterns, alongside personal wellbeing, value sets, and ethics.

In the generic tables (see Tables 4.1–4.4 below), you will be guided through each dimension, step by step, to discover how the process works. Each of the four dimensions is rendered as a simple chart containing a number of key drivers, for instance politics, communication, wellbeing, and quality of life. These headings are the overarching structural drivers likely to come into play in any discussion about the future, no matter what organization or business sector you operate in. The structural drivers are key components for creating a Trend Atlas template (Figure 4.1) and a vital starting point in any process of considering the influences likely to impact your organization in the short, medium or long term.

### SUMMARY: Your Trend Atlas

- Trend management is typically conducted in four phases – scanning and collection, analysis, synthesis, and communication.
- Initial soft analysis of data narrows down potential influences to a core group. Each macro trend, influenced by core micro trends, is fed into the Trend Atlas framework to create an initial map.
- Next, the process moves to the core analysis, leading on to a final synthesis of findings into a sector-specific people-centric universe, executive summary, and action plan.
- Findings are crystallized into a Trend Atlas (or consumer Mindset Map) for building viable future scenarios. This enables effective communication of findings through workshops and Q&A sessions to enable implementation.

## Identifying key trends in the multiple dimensions

Underneath each heading on the Trend Atlas framework (Tables 4.1–4.4), we have added a number of significant topics and influences that form the macro trends impacting societies, markets, and people in general. Any macro trend (or topic) placed in one of the generic diagrams is care-

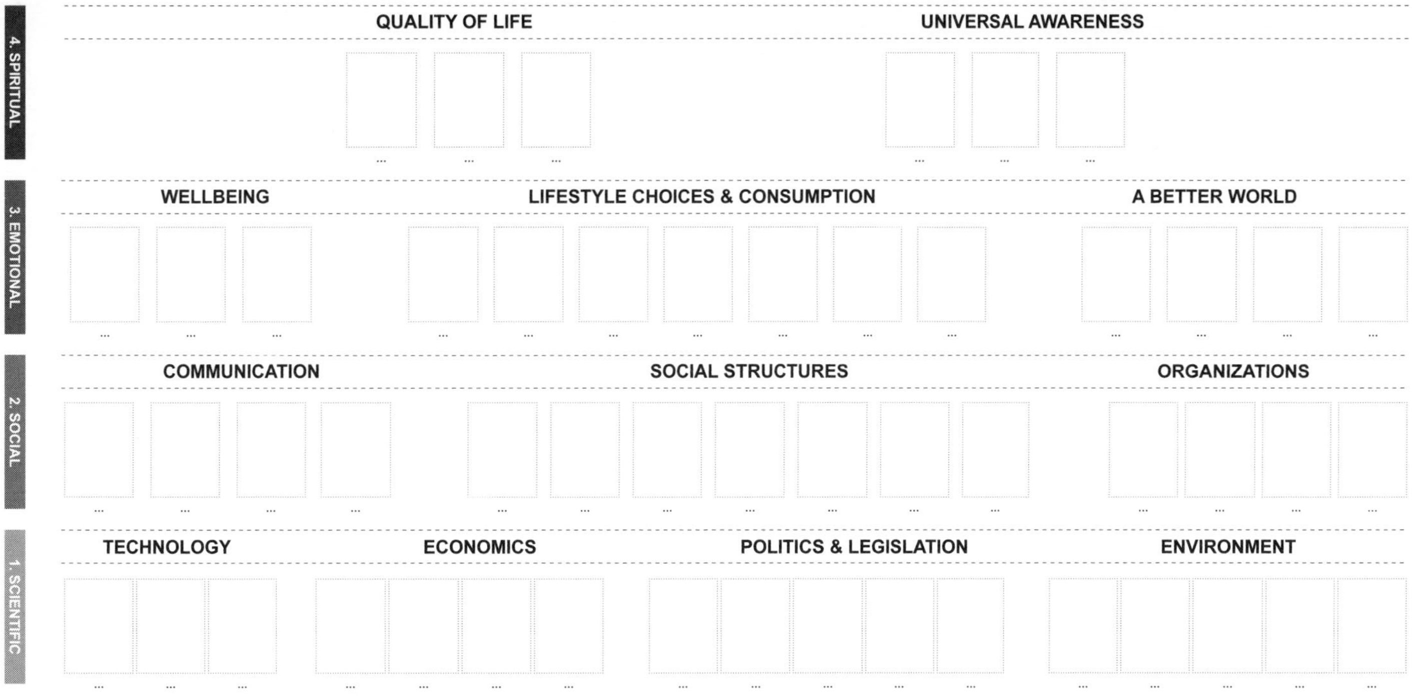

4. SPIRITUAL

QUALITY OF LIFE

UNIVERSAL AWARENESS

3. EMOTIONAL

WELLBEING

LIFESTYLE CHOICES & CONSUMPTION

A BETTER WORLD

2. SOCIAL

COMMUNICATION

SOCIAL STRUCTURES

ORGANIZATIONS

1. SCIENTIFIC

TECHNOLOGY

ECONOMICS

POLITICS & LEGISLATION

ENVIRONMENT

FIGURE 4.1 **Trend Atlas template**: The 4D template for mapping significant structural drivers and macro trends
*Source*: Kjaer Global

fully considered for its impact across the other dimensions, factoring in the natural interconnection that exists between the scientific, social, emotional, and spiritual domains. Note also that similar trends can appear across dimensions depending on their relevance and influence on a softer qualitative level or a more scientific quantifiable level.

It is important not to exclude topics in the initial discussion and selection of macro trends, as they might emerge as a key priority for your market or organization at a later stage of the trend-mapping process. Another key point to keep in mind is that this process is not prescriptive but collaborative, so topics can be added, amended or removed as you continue to drill down into your specific challenges and evolve your trend-mapping expertise.

The generic diagrams in this chapter provide a working template to inspire your own deliberations, and Tables 4.1–4.4 provide useful keywords/phrases within each of the four dimensions that might typically be considered within your own debates about the most relevant drivers and trends. In Chapters 5–7, we will apply the process in an in-action forecast of key trends to watch that will inform and shape our world, businesses, and organizations – and impact people's behavior up to 2030+.

## The scientific dimension

The scientific dimension (Table 4.1) is the foundation, and the most physical and concrete part of the Trend Atlas, where all the fundamental structural drivers that underpin our world and society as a whole are situated. Trends in the scientific dimension typically follow the traditional PESTEL (politics, economics, society, technology, environment, legislation) structure that many of us are familiar with, considering the large and quantifiable shifts in our surrounding environment. The society drivers remain constant and, although topical issues and new dynamics will emerge over time, change here tends to happen at a slower pace than in the other three dimensions.

### Scientific structural drivers and macro trends

**Politics:** Considers current and potential challenges within the landscape of public life. In the multidimensional model, it offers space to consider not only traditional political systems, but also disruptive or emergent forces. The breaking down of old power structures – inspiring everything from

TABLE 4.1 **The scientific framework**: Generic structural drivers and macro trends in the scientific dimension

| Politics | Economics | Society |
|---|---|---|
| ▪ Geopolitics<br>▪ Local/regional<br>▪ Equality and empowerment<br>▪ Political movements<br>▪ Accountability | ▪ New economies<br>▪ Global market turbulence<br>▪ Globalization<br>▪ Market opportunities/ threats<br>▪ New economic movements | ▪ Social challenges<br>▪ Migration and new communities<br>▪ Demographics<br>▪ Society frameworks<br>▪ Educational models |
| **Technology** | **Environment** | **Legislation** |
| ▪ Bioscience<br>▪ Nanotech<br>▪ Alternative energies<br>▪ Convergence<br>▪ Obsolescence<br>▪ Robotics<br>▪ Internet of Things (IoT) | ▪ Climate<br>▪ Resources<br>▪ Population<br>▪ Urbanization<br>▪ Environmental challenges<br>▪ Biodiversity | ▪ National laws<br>▪ Global governance<br>▪ Cyber policy<br>▪ Corporate regulation |

*Source:* Kjaer Global

new political movements to social media campaigns – is evidence of the growing demands for greater accountability from business and political leaders. This is especially pressing if we consider the potential challenges and opportunities that arise from discontent with current systems, where geopolitics is having a major influence in a local and global context. Meanwhile, the emergence of soft power pressure groups is challenging the way public life currently operates.

**Economics:** Examines the impact of evolving patterns, beliefs, and new models on the local, national, and world economy – especially in relation to how people use resources for the production, consumption, and distribution of wealth. It offers a forum to consider not only mainstream forces and academic theories, but also emerging value sets and aspirations that may challenge the status quo or the way we view prosperity in society. The economic systems that determine global and local markets are already a central theme in the 21st-century discussion on how we achieve sustainability and a more equitable system of living.

**Society:** Investigates the broader welfare issues that underpin our society, including changing demographics, health, education, and communities at large. It can be difficult to decide what to include or exclude from this category, since some components overlap and evolve into the social dimension of the Trend Atlas. While it is not too important to pin down where overlapping trends should fall in the initial discussions, it is useful to note that the scientific dimension often holds quantifiable trends on health, migration, literacy, and so on, whereas the social dimension focuses on behavioral and lifestyle patterns that emerge either because of these scientific trends or independent of them.

**Technology:** Investigates the effect of new and emerging life sciences and inventions, as well as exploring innovations that could revolutionize how we will live in the future. Technologies impact everything from health and wellbeing to architecture, manufacturing, work efficiency, and interface design, also playing a vital role in the global debate about reshaping society and business to create a better world. The adaptation of new technologies represents an opportunity but also challenges in many major industries operating today.

**Environment:** Encompasses key data and forecasts on local, national, and world issues such as climate change, resource shortage, biodiversity, and population growth. Scientists, urban planners, and designers, in collaboration with businesses, all have a key role to play in solving current and future environmental challenges. Increasingly, grassroots and citizen-led movements are driving change in these areas.

**Legislation:** Covers the effect of major laws and public policies, national and international, that exist or may emerge in areas ranging from emissions standards and wage levels to imports, taxation, health and safety, and transparency. This is a crucial area for any organization, since it encompasses all other markets where you may operate, now or in the future, whether directly or via the incorporation of raw materials/components and the use of services that originate elsewhere.

## The social dimension

The social dimension (Table 4.2) is concerned with people, organizations, and communication; it sits at the core of the Trend Atlas when looking at consumer trend management. Here, we explore the key social structural drivers with focus on behavior, lifestyle patterns, and emerging communication culture. There is some degree of overlap across the other dimensions, especially the emotional and spiritual, since our value universe is integral to how we live and the choices we make. The focus is on present weak signals that are signposting emerging trends, since most evolve and take on new meanings or foster other micro trends.

TABLE 4.2 **The social framework**: Generic structural drivers and macro trends in the social dimension

| Communication | Social structures | Organizations |
|---|---|---|
| ▪ Interconnectivity<br>▪ Sharing and co-creation<br>▪ Mobile technologies<br>▪ Global digital networks<br>▪ Dialogue-driven models | ▪ Identity and nationality<br>▪ Aging society<br>▪ Family structures<br>▪ Welfare and contribution<br>▪ Communities | ▪ New business models<br>▪ Innovation culture<br>▪ Workforce and talent<br>▪ Social capital<br>▪ Employment terms<br>▪ CSR and accountability<br>▪ Supply chain |

*Source:* Kjaer Global

## Social structural drivers and macro trends

**Communication:** Technology and growing interconnectivity are now an integrated part of our everyday lives. Mobile technologies are bringing a whole host of new elements into play – from co-creation and sharing to the explosion of hacking and gaming culture. The impact of mobile everything is rapidly changing the nature of education, finance, socializing, and work/leisure convergence, along with the challenges of private/public boundaries. In fact, this is one of the most important and fastest growing of all the generic structural drivers, touching all the other dimensions by offering opportunities to explore new disruptive markets and dialogue-driven innovation platforms.

**Social structures:** As we see a redefinition of the boundaries between youth/seniors, work/leisure, friendship/family, and social classes, structures are being redrawn. Mobility and digital interconnectivity mean we are developing new social networks and community structures – analogue and local or digital and global – as a result of a shared passions and affinities. This trend recognizes that traditional demographic segmentation, based solely on socioeconomic classifications, is no longer enough to observe behavior in an era when consumers are more situation-determined and community-focused than ever.

**Organizations:** Considers the ways in which changes in society affect the business landscape – including corporate social responsibility (CSR), disruptive business models, and new approaches to company structure and organization. It covers a vast range of subjects pertinent to business culture and social capital – ranging from human resources and empathic leadership to entrepreneurship and innovation culture as well as attracting and retaining talent. While the PESTEL factors, in particular, impact businesses within the social dimension, it is also important to examine in greater depth ways in which society is influencing how organizations will operate and contribute in the future.

## The emotional dimension

The emotional dimension (Table 4.3) investigates the qualitative lifestyle influences and is a vital area to consider because emotional values influence people's decision-making processes, needs, and wants, as well as their wellbeing and search for meaning. In general, people are looking for brand engagement and encounters that make them feel good about themselves, their future, and their contribution to society, so in this dimension, businesses should consider carefully how their existing corporate values match up to the trends they choose to explore.

TABLE 4.3 **The emotional framework**: Generic structural drivers and macro trends in the emotional dimension

| Wellbeing | Lifestyle choices | Consumption | A better world |
|---|---|---|---|
| ▪ Body capital<br>▪ Balanced living<br>▪ Foodie culture<br>▪ Health tech | ▪ Cultural capital<br>▪ Authenticity<br>▪ Engagement<br>▪ Inclusive design<br>▪ Storytelling<br>▪ Intelligent reduction | ▪ Collaboration<br>▪ New luxury<br>▪ Individualistic<br>▪ Experimental<br>▪ Meaningful exchange | ▪ One planet living<br>▪ Ethical consumption<br>▪ Collaboration<br>▪ Philanthropy |

*Source:* Kjaer Global

### Emotional structural drivers and macro trends

**Wellbeing:** Foodie culture and digital health to build body capital suggest a whole new preoccupation with living well for longer. While lifespan is generally improving throughout much of the Western world right now, a new crop of issues related to unhealthy and sedentary lifestyles are causing increasing concern. Living well for longer has sparked a whole new movement, in which people are guided to make the right choices.

This suggests a future in which education and support are required to help people make better choices. Organizations have a key opportunity in inspiring and helping people lead better lives – be it through inspiring awareness campaigns or by providing practical health solutions.

**Lifestyle choices:** An essential element in our perceptions about the world and how we see ourselves is how we live our life. With choice proliferating, different forces and trajectories are emerging – less traditional and linear than previously – they reshape lifestyle patterns and experiences. While our lifestyles are becoming more fluid, the search for deeper engagement will inform tomorrow's people, feeding directly into how we interact with organizations.

**Consumption:** Here we explore the way in which we use and enjoy products and services. As described in Chapter 2, overconsumption has become a big ethical dilemma with the growing availability of mass-produced goods, and a source of division between the prosperous and less prosperous in Western society. This makes it more likely that people will revisit their consumption choices, as growing unease about our impact on the planet inspires us to either consume less or choose alternative options.

**A better world:** This is closely linked to our consumption and lifestyle choices, and concerns people's desire to do the right thing and lead responsible lives – a clear challenge for organizations because higher standards of accountability are now demanded. Every organization must consider their impact on the planet and measure how they match up to demands for more sustainable and collaborative approaches to doing business.

## The spiritual dimension

The spiritual dimension (Table 4.4) sits at the heart of people's ethics and values. People are yearning for the "good life" and want to make a difference, and it is useful to refer back to the section Understanding the evolution of consumption in Chapter 2 in order to consider the history that informs these fundamental value drivers in a 21st-century context. Just two headings – quality of life and universal awareness – are listed in our generic table but, as previously stated, the process of creating a Trend Atlas is not prescriptive, so your organization may choose to frame the debate using different headings.

TABLE 4.4 **The spiritual framework**: Generic structural drivers and macro trends in the spiritual dimension

| Quality of life | Universal awareness |
|---|---|
| ▪ Mindfulness<br>▪ Self-improvement<br>▪ Belonging<br>▪ The "good life" | ▪ Active awareness<br>▪ Conscious capitalism<br>▪ Gross National Happiness<br>▪ Metaphysics |

*Source:* Kjaer Global

## Spiritual structural drivers and macro trends

**Quality of life:** Self-improvement – through better health/wellbeing and education – is a key motivator in today's society, while belonging (through work, interest, network or club) is especially important in a world where traditional social ties are loosening and many people move for their job or to achieve a better quality of life. This structural driver connects to our individual aspirations and actions to lead the "good life." Organizations with a purpose understand that it is part of their role to respond positively and enable people to achieve their goals.

**Universal awareness:** As individuals and organizations, our desire and need to make a difference within society is an area that increasingly informs our worldview and public debates. We see the impact in everything from social movements and "cause campaigns" to interactions between individuals and businesses. In this regard, it is crucial to consider what your organization contributes and how it might cement its role as a trusted and responsible player in society.

## Narrowing down the structural trends

After the overall trends are plotted, findings are filtered and mapped according to their categories and relevance. Typically, you consolidate and reduce your findings down to 10–12 structural drivers (headings), distributed equally throughout the four dimensions in the Trend Atlas. Each structural driver is then assigned a number of macro trends to kick-start the trend-mapping process. The best method for selecting, clustering, and aligning the core macro trends is a brainstorming session and this brings you to the final stage of creating your own Trend Atlas (Figure 4.2).

### SUMMARY: Identifying key trends

- The scientific dimension of the Trend Atlas follows the PESTEL model of politics, economics, society, technology, environment, and legislation. Change in this dimension happens at a slower pace.
- The social dimension is concerned with people, organizations, and communication. With some degree of overlap into other areas of the Trend Atlas, the focus is on weak signals that are signposting emerging trends.
- The emotional dimension investigates lifestyle influences that determine consumption, wellbeing, and search for meaning – highlighting how a organization's values match up to people's real needs and desires.
- The spiritual dimension explores motivations and values, fundamental to our desire for the "good life." It challenges us to consider future consumption patterns and how these will shape organizational strategy.

## The Trend Index

The initial process of plotting trends onto the Trend Atlas is designed to open up the debate and provide "fuel" for discussion about the future. In any study of medium- to long-term scientific, social, emotional, and spiritual society trends, it is important to frame objectives in terms of a holistic road map to the future. The next step is to consider the relationships between the key trends in all four dimensions – including overlaps and similarities – in order to narrow them down to those that matter most to your business.

### Next steps: best practice and methods

Choosing the most relevant trends for your organization and understanding them in a global and local context can initially feel like a real challenge, especially as most of us tend to see the world from our own cultural reference points and value universe. One option is to choose to work with the generic Trend Atlas (Figure 4.2) to select 8–12 core trends that are the most relevant and meaningful for your particular project or strategy work. The best way to do this is to generate a team consensus about what a shared future vision or goal might look like. By converting

4. SPIRITUAL

## QUALITY OF LIFE

Mindfulness

The Good Life

Happiness Hunting

3. EMOTIONAL

## WELLBEING

Active Leisure

Foodie Culture

Health = Wealth

## LIFESTYLE CHOIC

Authentic Storytelling

Cultural Consumption

Inclusive Design

Rea

2. SOCIAL

## COMMUNICATION

Education 4.0

Cloud Intelligence

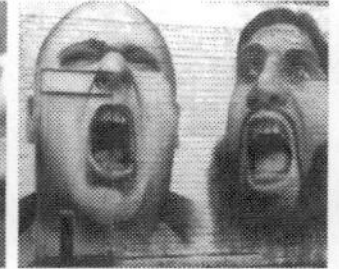
Open Dialogue

The Global Brain

## SOC

Redefined Families

Free-Range Parents

Female Factor

1. SCIENTIFIC

## TECHNOLOGY

Bio Revolution

Clean Tech

Thinking Cities

## ECONOMICS

Resource Shortage

Rising Economies

Turbulent Markets

Innovation Hubs

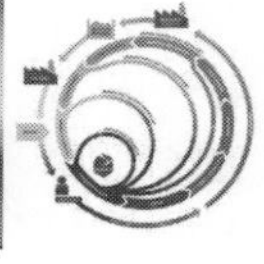
New Models

FIGURE 4.2 **Generic Trend Atlas**: The four dimensions with consolidated structural drivers and macro trends

*Source*: Kjaer Global

*Images*: New Models: "Mapping Design for Circularity," Credit: The Great Recovery Team at the RSA; Green Growth: Credit: Samsø Energy Academy

## UNIVERSAL AWARENESS

The Big Society

Enoughism

Purpose Driven Leadership

## CONSUMPTION

Collaborative Communities

Intelligent Reduction

Smart Living

## A BETTER WORLD

Betapreneurship

Good Causes

Considered Consumption

One Planet Living

## RUCTURES

Global Citizens

Creative Classes

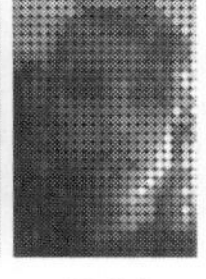

Digital Natives

## ORGANIZATIONS

Social Capital

Work/Life Balance

Lifelong Learning



Brand Engagement

## POLITICS & LEGISLATIONS

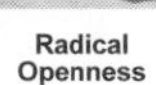

Radical Openness

Glocalization

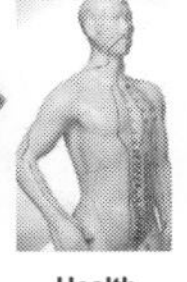

Health Challenges

Public Policy

## ENVIRONMENT

Hyper Urbanization

Crowded Planet

Green Growth

Biodiversity

Climate Change

your initial ideas into a more practical framework, you can start mapping your journey into the future. A collaborative workshop is recommended in order to engage your elected team of change-makers and give them the opportunity to take ownership over the process (Figure 4.3). Before the workshop, it's important to "frame the future" by inviting the participants to write a list of objectives for the workshop and include a time perspective for the changes you wish to consider over, say, 2, 5 or 10 years.

FIGURE 4.3 **Trend mapping in action**: Workshop participants work in groups, using a Trend Atlas to plot key macro trends for their organization
*Source*: Kjaer Global

Interaction-based methods used in multidimensional forecasting benefit significantly from being brought together with other disciplines and approaches. As already discussed, taking different perspectives into account provides a more accurate vision of the future. There are some tried-and-tested methods employed by many organizations; one of these is the traditional SWOT analysis. This tool is used to identify the main risks and opportunities impacting the organization, on an internal and external level. This method is the foundation of our Trend SWOT analysis (see Table 4.5), serving as a way to carefully explore the selected macro trends one by one. The insights generated from this exercise inform the foundation of your personal future road map.

## WHAT IS A TREND SWOT ANALYSIS?

A Trend SWOT analysis is a practical test to understand how best to act on your selected core macro trends. Exploring internal and external factors is the vital first stage in deciding what strategy to implement. Each of the selected macro trends must be processed in turn, addressing the core relevant questions that may influence your future, by using a SWOT diagram (Table 4.5). The internal drivers section focuses on strengths and weaknesses inside your organization, while external drivers explore opportunities and threats outside your control, including the market, the business climate, and competitors.

TABLE 4.5 **Trend SWOT analysis**: Internal and external key questions to be addressed

| INTERNAL DRIVERS | | EXTERNAL DRIVERS | |
|---|---|---|---|
| **Strengths** | **Weaknesses** | **Opportunities** | **Threats** |
| What organizational strengths do you have for tapping into this trend? | What organizational weaknesses prevent you from tapping into this trend? | What opportunities does this trend represent for your organization? | What threats could this trend represent for your organization? |

*Source:* Kjaer Global

This method, with its simple but central questions, frames and underpins the debate so that your team consider and plan for internal and external forces that affect your business in the short to medium term. The practical outcome is a future road map, with a structured implementation plan to support and drive your strategy.

Once the key macro trends have been carefully selected and analyzed, the next step is to create a Trend Index by mapping them. Trends must be observed as an ecosystem of potential change factors that represent risks and opportunities. In order to approach this, you should carefully consider how your brand, business model, people, products, and services are represented in your particular sector or organizational ecosystem.

Then you explore the micro trends supporting the individual macro trends one by one. The example Trend Index (Figure 4.4) shows you how to map macro trends with relevant supporting micro trends in a technology forecast. The interconnected macro trends set the framework on a higher, more general level in a global context. To explore them closely, in a specific context or sector, the "lower level" of more specific micro trends are needed. Typically, each individual macro trend is supported by between two and four micro trends, depending on the desired breadth and width of the lower level influences; naturally, the micro drivers will vary depending on your sector and industry. There may sometimes be more micro trends influencing a particular macro trend, but generally a maximum of four are recommended to reduce complexity.

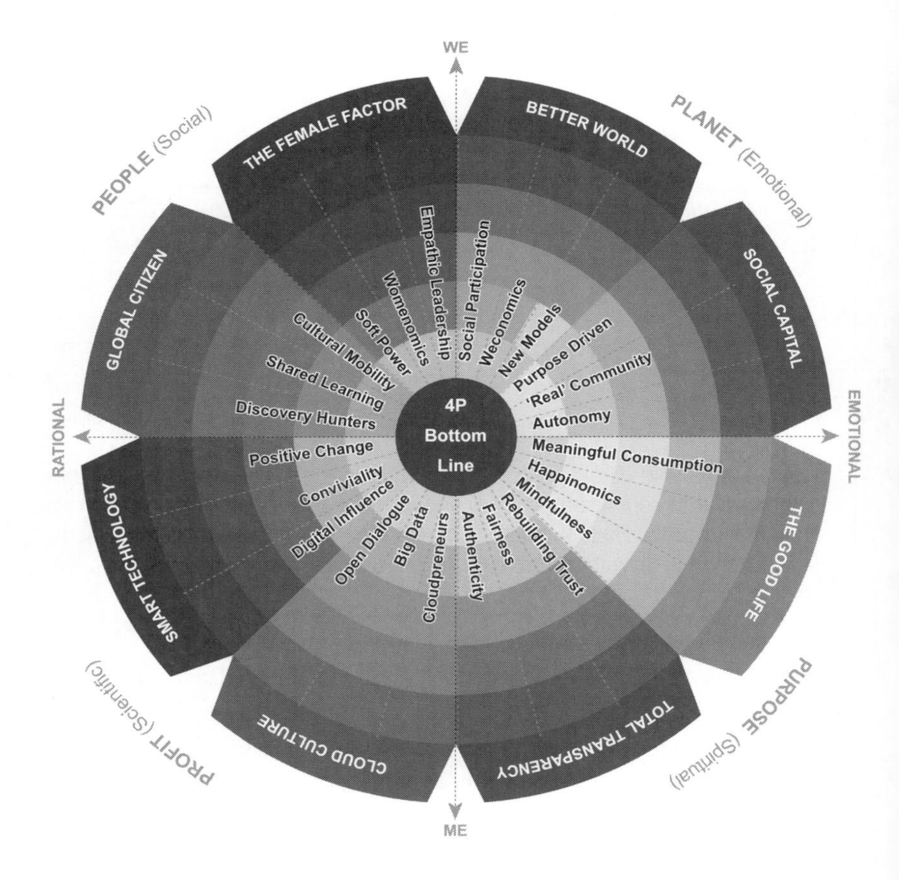


FIGURE 4.4 **Trend Index**: Trends are observed as an ecosystem of risks and opportunities
*Source*: Kjaer Global

## WHAT IS A TREND INDEX?

The Trend Index is an overview for bringing analytical simplicity to a complex set of data and insights from research, questionnaires or polls. It identifies the similarities and differences between the trends, as well as shared values and synergies within a multidimensional 4P framework of people, planet, purpose, profit. The index is a compass for assessing risk and opportunities in a global macro and a local micro context. Further, it is a useful benchmarking and trend relevance tool for strategic decision making.

The Trend Index has multiple uses, but first and foremost it provides a platform for understanding the greater context of the trends. In this regard, it can be used to assess trend relevance in a geographical or specific business context as it makes the trends comparable. Further, within a business context, it can monitor the trend performance in a time frame analysis, looking at implementation success and progress. Overall, it is a practical tool for collaboration and communication of trends and for revealing new business opportunities.

You cannot predict the way the future will play out with absolute certainty, but you can assess and monitor trends and consider likely future outcomes by mapping them in relation to each other; this also enables you to decide how to act timely on changes in your market, sector or society. Increasing complexity and accelerating change makes it vital to understand the interconnectedness of global trends to anticipate the forces at play in a local context. The Trend Index is a management tool for assessing risks and challenges, measuring trend engagement, making regional comparisons, and exploring local markets.

The Trend Index used as a Trend Engagement Barometer (see Figure 4.4) indicates values on a scale of 0–100%, with the highest scores represented by positions furthest from the center. Data can be gathered using qualitative or quantitative polls among internal or external stakeholders. Low scores represent challenges and possible risks, but they may also be weak signals of future opportunities. Higher scores indicate that the trend is more commonplace and perhaps already quite integrated within the mainstream. Rating the micro trends individually creates a more in-depth analysis of the macro trend's overall impact.

## CASE STUDY: READING THE BAROMETER IN THE FS MARKET

The Trend Engagement Barometer (Figure 4.5) was used to evaluate internal trend engagement within a large Nordic financial services (FS) organization in 2012. Top management and the team of change agents were asked to assess and rate the current implementation of the key macro trends. In a qualitative opinion poll, the group of internal change-makers responded to the question: How relevant are the trends currently looking overall at your organization, your people, and your customers? These were the results.

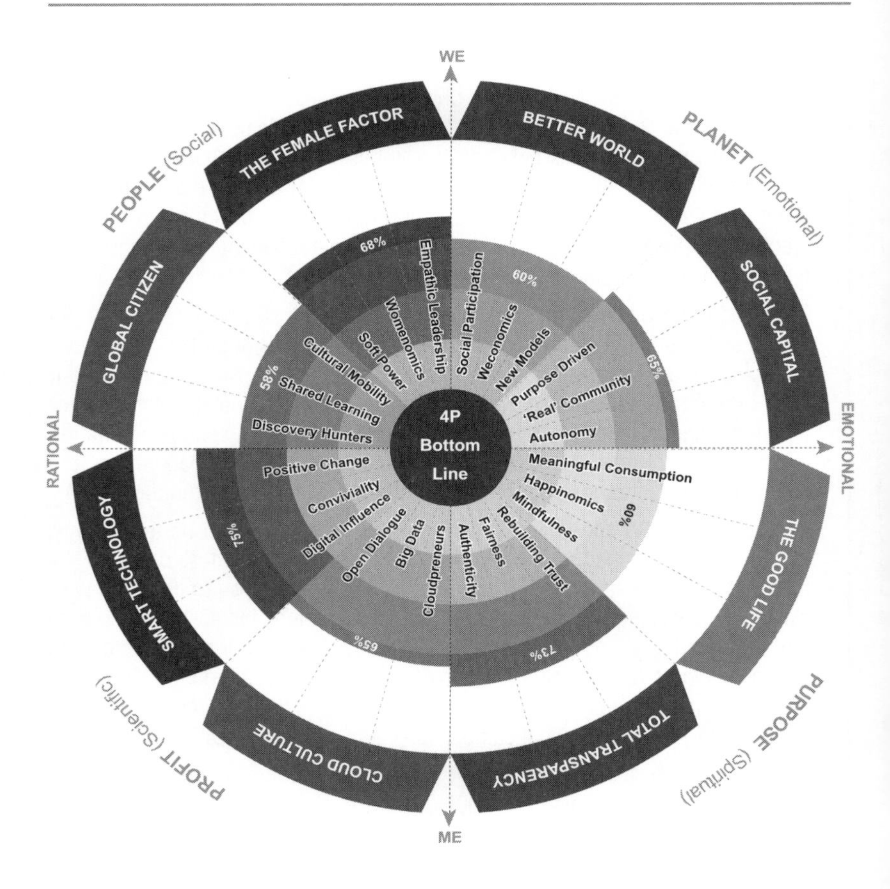


FIGURE 4.5 **Trend Engagement Barometer**: An example of a Trend Index measuring engagement
*Source*: Kjaer Global

**Profit:** Smart Technology was rated the highest of the trends, just below 3.8 (75%), indicating that the organization had already adopted new technology to drive growth. However, Cloud Culture rated at 3.25 (65%), suggesting more of a future challenge because of regulation and security within the FS sector.

**People:** The Global Citizen trend came lowest at 2.9 (58%), with a lack of a talent strategy to attract a younger workforce. The Female Factor at 3.4 (68%) indicated that women were a large part of the workforce, although not at top management level – leaving room for improvement.

**Planet:** A Better World only scored 3.0 (60%) as the organization's CSR strategy was being reviewed during this project. Social Capital at 3.25 (65%) indicated an opportunity to collaborate more closely with all stakeholders and to embrace community in order to grow Social Capital and Better World credentials.

**Purpose:** Total Transparency was rated at just over 3.6 (73%), indicating a commitment to openness and simplicity in products and service offerings. However, with the Good Life scoring 3.0 (60%), there was huge scope for using this trend as a storytelling tool to improve internal and external connections with people.

**Overall challenges and opportunities:** Initially, the rational values were rated highest, as would be expected in the FS sector. But it was evident that within the softer, more value-driven dimensions, there was both scope for improvement and untapped opportunities.

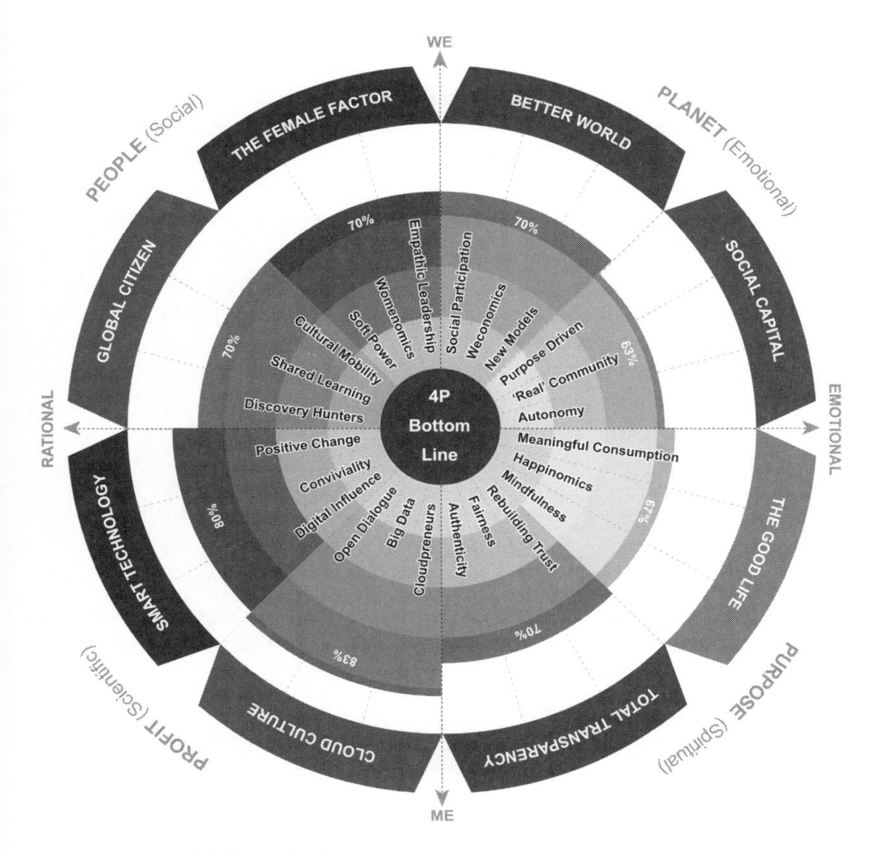


FIGURE 4.6 **Global Trend Relevance Index**
*Source*: Kjaer Global

## CASE STUDY: GLOBAL VS. LOCAL TREND RELEVANCE INDEX

The same organization was asked to rate the trend relevance in a general geographical context. The question: How relevant are the macro trends a) globally? b) locally? gave the following results (Figures 4.6 and 4.7).

**On a global level:** Technology was rated highest at around 80% and Social Capital came lowest at just over 63%. The remaining trends were all rated at around 70%, indicating a general consensus about the importance of these drivers. The people dimension, which focuses on new leadership models and talent, better world activities and transparency, is essential for actively promoting the "good life" and positive change (Figure 4.6).

**In a local context:** The Good Life and Female Factor both received a very high rating at 87%, positive news in a region where happiness and integration of women in the workplace are noted features. They came second only to Smart Technology at 4.5 (90%), while Cloud Culture scored 4.0 (80%) – possibly because of deep integration of technology within this region. Across Better World, Social Capital, and Total Transparency the average was well over 80% – suggesting transparency and social inclusion. However, Global Citizens was rated lowest (70%) – suggesting respondents saw a lack of opportunities to attract this valuable group to the local workforce (Figure 4.7).

**Vision 2020:** Comparing the global and local sample with the organizational Trend Index, it became clear that the organization should reappraise its objectives and work to reach a delivery score on the index of 4.5 by 2015 and 5.0 (100%) across all trends by 2020. The organization determined that a clearer focus on the softer values of Social Capital would attract and retain more digital-oriented young talent – including the all-important Global Citizens – drive innovation, and implement a more human-centric future vision to achieve overall company objectives.

## SUMMARY: Trend Index

- Plotting trends onto the Trend Atlas opens up debate about future directions. The next step is to consider relationships – including overlaps – between trends in all dimensions to narrow down the most relevant.
- One option is to work with the generic Trend Atlas and shortlist 8–12 trends most relevant to your organization and then hold a collaborative workshop to develop a strategy, setting a time perspective for changes you want to consider, for example 2, 5 or 10 years.
- A Trend SWOT (strengths, weaknesses, opportunities, threats) analysis is useful at the outset to identify risks and opportunities for your organization and underpin the debate.

- Once key macro trends are selected, they can be mapped onto a Trend Index to assess relevance in a geographical or specific business context. They should be observed in terms of risks and opportunities before each micro trend is explored. Each macro trend is typically supported by up to four micro trends.
- The Trend Index can be used as a Trend Engagement Barometer, a visual benchmark that indicates challenges and opportunities. Data can be gathered using qualitative or quantitative polls, among internal or external stakeholders. Higher scores indicate that a trend is more commonplace and mainstream.

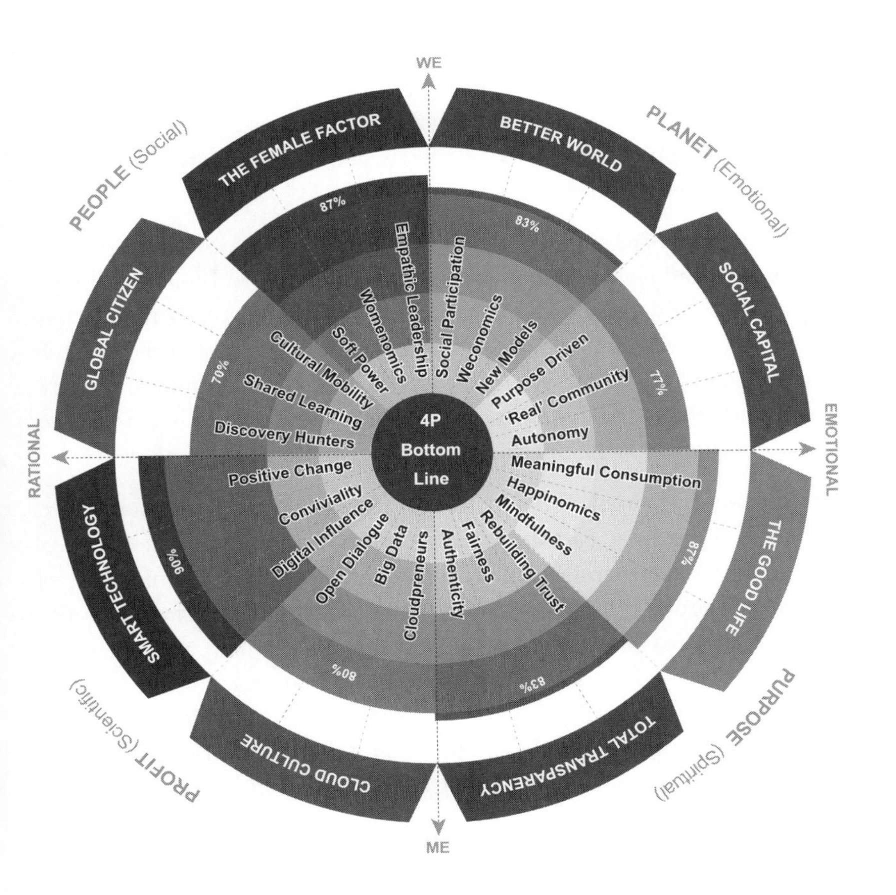


FIGURE 4.7 **Local Trend Relevance Index**
*Source*: Kjaer Global

## Adopting a glocal perspective

The last three centuries of rapid technological development and discovery have shifted the framework of ideas and beliefs, shaping the perspective through which we watch, interpret, and interact with the world, as organizations and as individuals. The big discourse of the 20th century was: What is shaping human behavior? Is it structure (the state and social influences) that restricts free choice and the opportunities to be independent or is it agency (the individual/ourselves)? The concept of reflexive modernity emerged in the 20th century to address these questions and was launched by sociologists Ulrich Beck, Anthony Giddens, and Scott Lash. Giddens argued that:[1] "The cognitive theory of reflexive modernization is optimistic at its core – more reflection, more experts, more science, more public sphere, more self-awareness and self-criticism will open up new and better possibilities for action."

Today, we are shaped less by society constructs and more by our immediate self-created reality and environment. But in a world inundated with information and contradictory messages, fresh thinking is required to adapt to a new world order. Organizations are an essential part of this discourse, and it is one in which they need to start telling new and heartfelt stories about the future to demonstrate their own role as reflective change-makers. But the question is: How are these stories emerging?

Michel Foucault's book *Order of Things*[2] touches on the concept of reflexivity. He argues that each historical epoch has an episteme or historical time frame that structures and organizes knowledge. The learning society is the philosophy of lifelong learning, one that is going to influence this century on a massive scale. The idea of lifelong learning and reflexivity is hardly new – the ancient Greeks believed that engaging in civic life and politics was a public duty. The problem here is that, in our pursuit of knowledge, as Foucault[2] puts it: "man is both knowing subject and the object of his own study." What is certain is that we need to frame the debate beyond our own experience and society perspective. Every individual worldview is important, but it is crucial to factor in how events and trends may manifest differently in various time fames and regions of the world.

## Our self-created reality and environment

What we know to be true of the 21st century is that the state is beginning to lose its significance as a supportive institution. With the rise of multinational corporations and organizations, the arrival of digital technology, the blurring of work/leisure, and the rise of social networks, people are reconsidering where they stand in relation to the society they inhabit. Traditional family ties are increasingly shifting towards more fluid associations; the power of conventional politics is evaporating as people lose faith in the values of their leaders; and for many, religion is no more than a cultural entity.

This brings us back to reflexivity. It has been argued for some time now that people no longer need to live by the values and "rules" set by outside forces (religion and state, employer and union). Instead, they have become the center of their own universe, where they choose to engage with people like themselves to define their own value parameters. Certainly, as we explore our world, physically and virtually, we are exposed to a much wider array of influences than ever before. The local impact of global knowledge sharing and the ways in which it alters perspectives is astonishing. Armed with digital tools and an active social network, it is perfectly possible to exchange ideas and stories with friends in Russia and South America; sign up to a petition in India; attend an environment webinar in Scandinavia; then scan a report on Asia-Pacific green growth and share it with business partners. Follow that with an Internet conference with the US and, within a few hours, you have been around the globe. So when we talk about glocal, what we mean is that, with easy access to truly local snapshots, we no longer need to rely on traditional media channels to filter and present insights.

## The century of the brain

Our worldview changes and evolves over time and it is important to recognize the profound and society-wide influences that shape organizational and individual perceptions. The 19th century was the "century of the nation state," one in which our reality was related to a geographical point of view. We traveled the world and found a whole new understanding of our ancient history and previous civilizations through archaeological exploration. The 20th century was the "century of the self," as Freud's psychoanalysis firmly planted the first seeds of self-discovery. Within this universe, we gained awareness of how our social code could be altered by physical goods and we became empowered consumers. Still in its infancy, the 21st century is shaping up to be the "century of the brain," as we witness a new era of neuroscience progressing at a fast pace to discover the underlying mechanism of behavior and diseases. In tandem, our digital reality is one in which devices, people, businesses, and public institutions all connect through an interconnected web of the Internet of Things (IoT) – a massive collective Global Brain. Already, many of us do not store things physically or locally, but rather digitally and globally. This suggests a prevailing influence of knowledge, in which we need less physical space but more memory and collective access to information and empowerment. The potential of a global digital reality is vast – with its emphasis on open-source learning, sharing, and collaboration – as a tool to solve many of the world's great challenges. It seems certain the "century of the brain" will reshape the way we think, feel, and explore our world (Table 4.6).

TABLE 4.6 **Historical time frame**

| HISTORICAL TIME FRAME – INFLUENCES SHAPING SOCIETY | | |
|---|---|---|
| **19th century** | **20th century** | **21st century** |
| Industrial Age | Technology Age | Multidimensional Age |
| Century of Nation State | Century of Self | Century of the Brain |
| Physiography | Psychology | Neuroscience |
| Phrenology Approach | Split-Brain Approach | Global-Brain Approach |
| Royal Geographical Society (1830) | British Psychological Society (1901) | Internet of Everything (IoE) (1999) |

*Source:* Kjaer Global

## Looking from the outside in

Cultivating a culture of authenticity and community is a primary means to build trust and business performance. The process of understanding how markets and people perceive your organization is vital, whether as a brand, a product or service provider, because converging forces are transforming relationships and the main agents in society. In a scientific and social context, technology convergence has empowered our working and personal lives. And in an emotional and spiritual context, we see opportunities for companies to create personal meaning and enablement – to show that they care. This, in a nutshell, is why it is so important for organizations to create a meaningful map of the future.

## Your GPS to the future

When we compare how companies operated two decades ago with today's situation, it is evident that a fast world with increasingly shorter lead times has left many businesses with very little time to think in the long term. Rather than plan ahead, they operate in the here and now and miss valuable strategic thinking time that a considered step-by-step process can offer. Therefore, a logical reflective framework for navigating the future is essential to create an authentic map of experiences – your individual company road map to facilitate your journey ahead.

Looking from the outside in to consider how your organization appears to the outside world is the starting point of the process of developing future strategies. It's also important to see the bigger picture – internal and external forces – via a brainstorming session or a workshop to start the debate, open up minds, and begin the process of defining and sharing a future vision alongside initial achievable milestones. At its simplest, the debate begins with three questions: Where are we now? Where do we want to go? What is preventing us from getting there? After carefully examining internal and external forces using a SWOT analysis, participants write down their initial vision and thoughts in note form. What is most essential is to map out your objectives on a short-, medium- and long-term basis – engaging in an iterative process, where you reflect on and adjust your road map to the future as you go along.

## Fostering innovation within your organization

While the generic Trend Atlas is a useful starting point, it is important to recognize that it does not provide you with a set of made-to-measure answers, as every organization has its unique brand DNA. You still need your individual judgment in order to make informed decisions about what criteria, parameters, and conditions to include and exclude. Building your own Trend Atlas from scratch requires time, but it is a strategic document that can be used to inform and shape your innovation and business strategy over a decade or more. While the most basic questions that frame the strategy process deal with direction, it is also important to ask and answer the larger question: What is our dream scenario for tomorrow? In order to foster innovation and develop a vision, we cannot restrict the frame of future strategy within the safe remit of what we have already done or know we can safely achieve. Having a vision – a dream scenario – is not about taking crazy risks, but planting ideas to cultivate the seeds of innovation culture to enable intrapreneurship to thrive within our organization.

## Creativity is like jamming

In a competitive environment, innovation culture is a crucial distinguishing feature of successful companies, but it needs to be cultivated in the right conditions to blossom and grow. In his book *Jamming: The Art and Discipline of Business Creativity,*[3] John Kao explores how companies that understand and foster creativity within their organization succeed. Kao demystifies creativity, a topic that many business leaders find difficult to approach or implement with confidence, by approaching it from the perspective of jazz, and defining a vocabulary and a grammar that can be observed, analyzed, understood, and taught. He argues that, like musicians in a jam session, business leaders can learn to stimulate creativity within their organization by enabling and encouraging people to feed off each other's ideas and so develop something new.[3]

In Chapters 5–7, we set out a range of perspectives on the future to assist your organization in creating a sound future road map. We dive into the key society drivers beyond 2030+ – our shortlist of the ones to watch – with key insights and case studies to assist you in understanding and mapping your industry or sector and consider consumer perspectives. With an increasing demand to deliver products and services solutions that capture the zeitgeist and match up to our real needs, we also explore how exponential technology impacts innovation on all levels in society and how new value barometers influence reputation and brand building. Used together, Chapters 5–7 will enhance your understanding of how to navigate the future by providing the tools, insights, and case studies needed to kick-start your organization's journey into the future.

## SUMMARY: From the outside in

- People's traditional family ties are shifting towards more fluid associations and, in a world inundated with information, organizations must start telling new and heartfelt stories about their own role as change-makers.
- Cultivating a culture of authenticity and community is the primary way organizations build trust and business performance. This is why it is so important to create a map of the future.
- This means looking from the outside in to consider how your organization appears to the outside world. The strategy debate starts with the process of considering where you are now and where you want to be, but it also invites you to consider your dream scenario of the future.
- Fostering an innovation culture means encouraging creativity and developing intrapreneurship programmes, encouraging people to feed off each other's ideas to develop and foster new thinking.
- Every organization has its unique brand DNA and, while building a Trend Atlas from scratch requires time, the resulting document can be used to inform and shape strategy for a decade or more.

## Chapter 5

# Major trends to 2030+

*Trends are not static but organic influences and elements that interact to shape and reshape our world. No organization can ignore their importance in shaping tomorrow's socioeconomic and cultural agenda.*

To observe trends in any setting and then incorporate them into a leadership, R&D, design or innovation strategy, it is crucial to understand how they are interlinked. The difference between having information and being able to turn it into tangible, meaningful knowledge is one of the greatest challenges for anyone working in the medium to long term, thus a knowledge-filtering platform is crucial. This is designed to allow for the best ideas, data, and qualitative insights to rise to the surface, either because they stand out as inspiring or because similar themes recur in a way that indicates their importance to future planning. Once the key influences are selected, you have a viable springboard to assess likely developments, strategies, and innovations.

## Major structural drivers

By 2020, Millennials will form half of the global workforce, and this group will have a very different view of what constitutes "progress."[1] Meanwhile, the aging population is already influencing health forecasts, with the World Health Organization (WHO) predicting that dementia cases could reach well over 65 million by 2030.[2] It is also forecast that diabetes will affect over 590 million people globally within 25 years, with an "epidemic" in low and middle income nations,[3] while over 42% of

Americans will be obese by 2030.[4] The growth of diagnostic technologies and increasing health awareness could potentially result in a more positive future outlook, but with these staggering numbers in mind, we clearly need to revisit current lifestyle preferences in order to map out balanced recommendations for living well for longer.

The UN predicts that we will be 8 billion people by 2025 and 60% will be living in urbanized areas.[5] By 2030, more than 5 billion of us will be online – the majority being in Asia, South America, Africa, and Russia.[6] Internet connectivity is set to rise exponentially, especially throughout Asia, and McKinsey has forecast that China and India will be the world's largest Internet users by 2015, followed by the US.[7] These vast human networks will reshape how we view the world and interact, bringing a new form of democracy and greatly increasing our demands for radical openness from those who govern us. Smart organizations will be those building bridges rather than walls by embracing the values of collaboration and openness. According to Edelman's 2012 Trust Barometer, already 65% of people globally said that transparent and honest business practices are the key to corporate reputation and trust.[8] If we fast-forward to 2030+, the world will perhaps not look very different to today, but it will be vastly more interconnected, transparent, and "smart" – meaning that better standards become a given in all areas of life.

## A people-centric vision

The trends in this chapter all derive from the socioeconomic and cultural drivers influencing today's lifestyle patterns while shaping tomorrow's society. Thinking cities and a smart society will be a reality that interlinks government, businesses, and people alike in an ecosystem of actions and reactions digitally connected to a vast Global Brain. Together and right now, these are shaping new cultural paradigms that will have a far-reaching impact. The Internet of Things (IoT) – also called the Internet of Everything (IoE) – is evolving faster than ever, and with Cisco projecting that as many as 50 billion devices will be connected by 2020,[9] growth and potential in this area are simply immense. The IoE enables enormous opportunities in all areas of life – politics, education, media, health, commerce, and leisure – add people to the equation and you have the foundation for a true smart society.

While exponential digital technology, faster lives, and the radical transformation of social mores have empowered humanity, they have also left many feeling unbalanced because they are too connected and too reliant on products and services that didn't even exist not so long ago. However, the benefits of technology are indisputable, with advances in areas ranging from education and healthcare to manufacturing and sustainability – all recent developments that will solve many of the world's future challenges. Still, there remain concerns that progressive technologies are actually detracting from rather than enriching lives. For instance, some question if the loss of physical meeting points and serendipity because of social networking is actually making us more insular and lonely. No doubt every era has its plausible experts warning us about the downsides of technology, but whatever our individual perspective on this issue, we cannot deny that technology is also a global engine of economic growth. It is in our power and self-interest to turn tech innovations into an advantage that will benefit and influence multiple levels of society; in order to do so, we must recognize that although the world has become digital, people remain analogue.

We increasingly expect products and services to deliver social cohesion, optimization, and general wellbeing alongside material benefits, and the future will be, as Havas Media CEO David Jones has predicted: "who cares wins."[10] Ultimately, the smart society is an inclusive vision where everyone can be change-makers and active participants in shaping the future they want to see. So, whether you are in top management, R&D, innovation or education, "made for people" will be the guiding principle; once you have included the human dimension, you can actively safeguard that your solutions are inclusive and beneficial so that people and business can thrive together.

## The ones to watch

The Trend Atlas is crucial for navigating complexity, as it invites businesses and organizations to think "from the outside in" to understand and connect with their audience, whether partners, employees or customers. From the generic Trend Atlas 2030+ (Figure 4.2), we have selected 12 of the most influential trends – the ones to watch – in order to define how they may develop and shape the next 10–20 years on a grand scale. These are major society-wide trends, already influencing society, business practice, and consumer behavior. No one can predict exactly what the future will look like, but by attempting to understand the anatomy of current society influences and early "weak signals," we assemble the building blocks that enable us to be well prepared and to embrace the future potential that change brings. Our outlines serve as an inspiration for any future strategy debates and an example of how mapping trends works in practice.

## How to read and understand the trends

As explored in Chapter 4, the scientific and social dimensions – containing the PESTEL-influenced drivers – are more static and enable us to observe and monitor trends in the medium to long term, whereas the emotional and spiritual dimensions are the human-centric elements that will evolve and take on new meaning over the next few decades. We have chosen to investigate the trends in a wider panoramic context, rather than just from a narrower consumer point of view. This panoramic method explores the multiple layers needed to gain a nuanced insight into the influences shaping our future reality. However, all the selected structural drivers and supporting macro trends also underpin the evolution of people's behavior, lifestyles, and consumption patterns. While we have explored each of the trends individually, it is important to remember they are closely interlinked.

The society drivers and macro trends pinpoint, in brief sound-bite form, where change is currently happening and who is pioneering new thinking in this context. They are a result of detailed research that includes the analysis of current trends, based on the framework and system outlined in Chapter 4 and underpinned by desktop research, wide-ranging literature reviews, academic papers, and articles from multiple sources. Evidence to support these trends has been provided throughout, along with real business cases or challenges explored in our workshops, round tables, interviews, and future opinion polls with clients and experts.

Each section starts by defining the trend (What) and then elaborates by exploring current macro influences that drive this change. The case studies illustrate early adopters (Who) implementing the trend, and we then consider emerging and evolved geographical areas (Where) this is happening. Finally, we outline the business case (Why) for the trend by considering it under the 4P cornerstones of people, planet, purpose, and profit.

### Trends versus time

While the trends are – as you would expect – illustrated in a more general way, as a reader you are invited to explore the specific insights for your business, also considering the likely risks and opportunities this may represent in your environment. The How to Implement section for each dimension provides a useful recap of the key vision for each trend, with guiding action points to enable you to consider practical applications in your business setting.

The selected trends represent the most interesting current insights and beliefs and, inevitably, they might be less pressing issues in, say, one or two decades provided we start to act on them. But in the current climate, they represent distinct challenges or opportunities that should not be ignored by anyone wishing to capitalize on change. This is perhaps also one of the best arguments for the relevance and purpose of a good Trend Management Toolkit; it is a call for action that inspires deep consideration of the issues we should not ignore in our business planning. While projections extrapolated from the information we have now may well shift and evolve over time, this does not reduce the value of developing strong narratives in the present about the future. Reading and interpreting signs in an intelligent and informed way is our best platform for developing visionary strategies and becoming true change-makers.

## Trends in the scientific dimension 2030+

## 1. Radical Openness

### What: Building bridges not walls

Authentic and trusted organizations ensure that brand promise and consumer experience are totally aligned. But one of the great challenges in a diverse and multicultural society is to maintain reputation at every level, from stakeholders through to industry sector and government. This challenge is growing in tandem with the digital economy, meaning businesses need to work harder – not only to be noticed, but also to be trusted. The key to gaining this trust is the adoption of radical openness, in which organizations enable and encourage scrutiny and public comment about their affairs. In short, smart companies will be building bridges, not walls, in the coming decades.

### Fostering total transparency

The Internet and digital technologies have raised expectations of personalized information and services on demand – empowering people more than ever before – and this is only set to increase in the future. Trust is earned through openness and consistent performance, so organizations must continually strive for value-creating solutions and experiences. Already, many people choose to bypass institutions if they appear opaque and inauthentic, so the challenge will be to align stakeholder interests with business goals in order to foster cohesion, inside and outside your organization.

### Digital reputation management

Improving your public image is not just about allocating more funds, but revisiting the underlying principles of your business model and making sure they sustain your organization and benefit all stakeholders, including the wider society. Using established digital platforms to communicate your message will ensure ease of access and transparent dialogue and it is a rewarding way to build and maintain relationships. Social media has already caused a global paradigm shift, influencing people's habits and behavior, and this presents opportunities to not only utilize existing vehicles, but also to develop company interaction platforms alongside social networking.

## Who: Scalable transparency

Google has become a global giant, indicating that transparency can work among even the largest organizations. The search engine is renowned for continuing to host the kind of weekly meetings it ran in its earliest days. Staff are still encouraged to ask questions of senior leaders – including founders Larry Page and Sergey Brin – an approach described as evidence of the innate transparency in the Google senior leadership team's DNA.[11]

## Where: The Nordic Way – open source, open minds

According to the World Values Survey,[12] which monitors long-term value shifts in over 100 countries, the Nordics are the world's biggest believers in individual autonomy, regarding the state's main job as promoting individual freedom and social mobility. Their tradition of governance emphasizes consensus. For example, in Norway and Sweden, people's salaries are publicized via the Internet, and the OECD assessed Denmark – where citizens' participation in the political process is extremely high – as having the lowest corruption rate in the world. This may explain why high levels of trust in government exist in Denmark and throughout the region. The report, *The Nordic Way*, concludes:

> Even if there is very little in the Nordic historical experience that is transferable to other cultures, it does bring one important point to the discussion: economic policies that cater both to our desire for individual autonomy and our need of community and security can be remarkably successful.[13]

## WHY: RADICAL OPENNESS BUSINESS CASE

### People: Earning respect

**Be an authentic employer:** A global survey of 97,000 people in 30 countries found that almost half (48%) would not recommend their current employer,[14] suggesting that new accountability standards are needed in order to attract and retain talent and be valued as an organization.

**Learn to listen:** With open communications, it is crucial to listen to what people say. Carl Bildt, Sweden's former prime minister, and a champion of the Twittersphere, suggested that it's about "getting to the pulse of what's happening." This will become crucial, not only for reputation protection, but also respectful relationships with all stakeholders.[15]

### Planet: Leadership and ethics

**Make a stand:** Strong leadership sometimes involves making a stand by taking tough decisions to drive an ethical agenda, but research suggests that we need more support from strong governments to solve global challenges like climate change.[16]

**Sharing principles:** Movements underpinned by equitable principles – notably the cooperative movement – look set to grow to match people's focus on ethics.[17]

### Purpose: Brand building and trust

**Fight corruption:** According to Transparency International's annual global survey, almost 80% of businesspeople surveyed believe their organization has an ethical duty to fight corruption.[18]

**Focus on brand DNA:** The financial crisis showed how whole industries can be destabilized by events. Financial services brands lost their trusted status, highlighting how important it is for all industries to focus on and share the what, how, and why of their brand DNA.

### Profit: Corporate transparency

**Transparent profit:** Openness helps to engage the workforce and means that everyone from the top level down to the most junior employees is more likely to feel accountable in their role and responsive to management needs when big changes are necessary.[19]

**Being real:** Writing in *Forbes*, John Hall argued for losing the "BS" in business and being "real." He cites a list of successful companies doing just that with bold initiatives such as pinning financial data on the office wall for all to see. Other examples include the CEO who shared his performance review and publishes company failures so other organizations can learn from his mistakes.[20]

## 2. The Global Brain

### What: Everything is connected

The Global Brain is a future vision of a universal, intelligent information-processing system that will connect data and information worldwide. Already, the web is altering public–private boundaries and ultimately this Global Brain – or the IoE – will link people, businesses, government, and our environment. In the process, it will reshape our sense of home and our definition of identity and community – the so-called "third space." This has profound implications for where and how we live and interact, so understanding cities and our environment as living organisms will be critical for developing intelligent systems that talk back to us.

#### Urbanization and thinking cities

The WHO has forecast that 60% of people will live in cities by 2030, and this is predicted to increase to 70% by 2050.[21] More than 20 of the top 50 largest cities by GDP will be in Asia by 2025 (as opposed to 8 in 2007)[22] and the region will become a crucial Internet hub, with China and India hosting the world's two largest online communities.[7] Urbanization has already created an urgent need for smart, connected environments that capitalize on advances in data analysis, sensor technologies, and urban design. Thinking cities are set to become centers for the strategic integration of IoE, using Circular Economy (CE) principles to deliver efficient self-supporting systems, alongside better infrastructure to benefit nations, businesses, people, and our environment.

#### Consumer dialogue and understanding

The increased ability to store, monitor, and measure data reaches into people's personal spheres, quantifying and analyzing every aspect of our lives. Responsive consumer dialogue is now central to business practice, as well as government. Huge data sets will provide marketers and planners with a much deeper and more strategic understanding of people, but questions are already being asked about their rights to hold and use our personal information. In this context, the Global Brain will continue to raise issues and spark worldwide debates about data protection. This is why it is essential to corporate trust and for all organizations to be transparent about the information they hold and its security.

### Who: Manufacturing systems

The Industrial Internet is taking off in a big way for manufacturing and this is already saving businesses vital money, time, and resources. It has the potential to connect people, data, and machines in a multilayered network that may add $10–$15 trillion to global GDP over the next 20 years. For instance, GE's Intelligent Platforms concept is envisaged as operating across sectors as diverse as aviation, homes, and hospitals to diagnose and fix products and parts before whole systems break down.[23]

### Where: New York City big data statistics

Big data enables organizations to quickly scan vast pools of information in their networks that might contain the answers to crucial challenges in our environment. New York City's Office of Policy and Strategic Planning is a "geek" team that analyzes city data for solutions to difficult and costly urban problems. An investment of $1 million has paid off brilliantly; the in-house team has leveraged the power of big data to tackle environmental dumping, cut fraud, clear hurricane debris, assist housing inspectors target fire-risk rental properties, and crack down on stores selling bootleg cigarettes. "There's a deep, deep relationship between New Yorkers and their government, and that relationship is captured in the data." [24]

## WHY: GLOBAL BRAIN BUSINESS CASE

### People: Connected ecosystems

**Saving lives:** Google was able to track where the H1N1 flu virus was likely to strike next in the US by taking the top 50 million keywords Americans were typing into search bars and comparing them with regional health statistics.[25]

**Eyewitness influence:** Ushahidi, a software platform (the name means "bearing witness" in Swahili), maps and analyzes information using the big data model, enabling people on the ground in disaster zones and political hotspots to be heard. Projects so far include eyewitness mapping of the Haiti earthquake and post-election violence in Kenya.[26]

### Planet: Circular thinking

**The city as a network:** Since cities consume 70% of global energy, sustainability starts here.[27] Examples of this circular thinking include Linköping, a city in Sweden, and its use of waste biogas to power public transport and Singapore's advanced water treatment and reuse systems.

**Integrated architecture:** Towards 2050, experimental city architecture will have developed advanced smart solutions to the challenges of building shortage and underuse; for example, changing the current classification of buildings into residential and commercial to enable smarter use of spaces on an "on-demand" basis.

### Purpose: Joining forces

**Acting as one:** Local government doesn't drive development in smart cities, but facilitates an "ecosystem" to manage investment across departments.[28]

**Environmental benefits:** The US National Ecological Observatory Network aggregates information on the ecological health of the nation, including tracking climate and land use change and invasive species.[29]

### Profit: Interconnected operations

**Data needs people and process:** IoE's ability to combine data with people, processes, and things will provide competitive advantage for companies that harness its capabilities.

**Technological value:** By 2025, disruptive technologies – including mobile Internet, automation, and IoE – could deliver economic value up to $33 trillion a year worldwide.[30]

## 3. Green Growth

### What: Collective considered consumption

Working in partnership with nature is essential if our cities are going to be resilient to population growth and climate change. By 2030, green infrastructure will not be "nice to have" but "necessary to thrive"– and this makes investment crucial. But current debates on sustainability are in need of a serious makeover, only 28% of US consumers know what terms like "environmentally friendly" and "green" actually mean. More alarmingly, only 44% trust green claims from big brands.[31] On the plus side, in recent years sustainability has moved from a philanthropic and CSR goal towards a real business opportunity. The number of companies that profited from sustainability climbed from 23% to 37% in 2012.[32]

#### A thriving Circular Economy

The CE is a vision to optimize resources and minimize waste by including reuse and renewal in the manufacturing process to deliver more sustainable products and services. Industrial ecology and regional eco-cities are the main solutions to implement the CE. Today, only 10% of global waste is recycled, but China has adopted the CE as it offers rich potential – its recycling industry is projected to be worth £183 billion by 2015.[33] It is estimated that UK businesses might benefit by up to £23 billion per year by 2020, through low-cost or no-cost improvements in resource use efficiency. Globally, total materials cost savings for industry could eventually reach over US$1 trillion per annum by 2025.[34]

#### Cleantech is the new gold

Clean technology (cleantech) has experienced a gradual integration globally, with a handful of nations leading this trend, notably Denmark, followed by Israel, Sweden, Finland, and the US.[35] Forecasts suggest that by 2018 the three main cleantech sectors of solar photovoltaics, wind power, and biofuel will have revenues of over $325 billion.[36] The rapid growth of renewables has impacted utility prices and undermined current business models across much of Europe's energy sector – most notably in Germany – so current and new players will have to embrace the renewable revolution.[37] By 2050, Europe will be interconnected by a smart electricity grid that distributes energy along gigantic underground power "motorways."[38]

## Who: Green growth initiatives

GE is said to be the company with most working relationships within the Global Cleantech 100, followed by Waste Management, Siemens, Google, and IBM.[39] It has been estimated that the CE represents potential material cost savings of $380–$630 billion annually within EU manufacturing sectors.[40] However, more needs to be done to develop the full potential of this sector. In an opinion poll at Ernst & Young's risk conference in Stockholm in 2013, with a focus on sustainability, Kjaer Global asked the audience if a CE would take sustainability to the next level. Over half agreed that this would be key to our future sustainability agenda; however, 38% believed that policies must be put in place to enable this to happen.

## Where: The world's green economies

**Asia** has become the cleantech laboratory of the world for many reasons. First, it is challenged by urban pollution, and second, its existing resource shortage has been viewed from the perspective of economic growth opportunities. To date, China has boosted green energy investment more than any other nation in the world.

**Denmark** aims to run entirely on renewable energy by 2050; it is already a global leader in creating green growth and jobs in energy efficiency, renewables, water treatment, and recycling. A prominent example of its forward-thinking approach was the 1997 Danish Ministry of the Environment competition to inspire local green growth. The winner was the small island of Samsø, with only 4,000 inhabitants, for its 'bottom-up' greening strategy to become Denmark's model renewable energy community.[41] Its entire energy needs come from sources like wind, sun, and biomass, and it is the world's first island to be 100% powered by renewables, making it a world leader in sustainability. In 2007, Samsø opened its own Energy Academy, and each year it arranges learning workshops and exhibitions that attract more than 4,000 politicians, journalists, and students on fact-finding missions. Søren Hermansen, its founding director, was named among the world's top 100 thinkers by *Time* magazine in 2008. Samsø's long-term goal is to be a fossil-free island by 2030.

**Israel** has an entrepreneurial culture and technical know-how that make it a natural hub for cleantech innovation. It is particularly strong in water and agricultural innovations, driven by its local climate and water challenges. A supportive investor and incubator base assists young companies.[39]

## WHY: GREEN GROWTH BUSINESS CASE

### People: Green job opportunities

Common causes: With a clear benefit in terms of sustainability, innovation, and competitive advantage, green job opportunities also give local communities a common purpose.

CE employment: Industrialized economies are seeking to reinvent their manufacturing models to create resilient domestic employment.[34]

### Planet: Alternative energies

Resource shortages: Mainstream projections suggest continued demand growth for major resources – from fossil fuels to food, minerals, fertilizers, and timber – until at least 2030.

Renewables: The WWF (World Wide Fund for Nature) has forecast that 95% of global energy demand could be met by renewables by 2050, given the right combination of investment and strategy.[42]

### Purpose: Sustainable development

Green road maps: The EU has agreed to address raw material security via its Roadmap to a Resource Efficient Europe strategy.

Reimagined cities: The organization Global Green USA has influenced more than $20 billion of green building initiatives. Notable projects include reimagining New Orleans through the Build it Back Green development project in the wake of Hurricane Katrina.[43]

### Profit: Green business initiatives

New business models: Nearly 50% of companies have changed their business models as a result of sustainability opportunities.[32]

**Green markets for growth:** In Germany, revenues from environmental industries have been predicted to more than double from 2009 to 2020, amounting to over €3 billion by 2020.[44]

## 4. Rising Economies

### What: Dynamics of a new global middle class

Described by the UN as a historic shift not seen for 150 years, the new middle classes in China, India, and Brazil have boosted their economies to equal the size of the G7 countries. Forecasts suggest that by 2050 they will account for nearly half of world manufacturing output.[45] By 2025, nearly two-thirds of the world's population will live in Asia,[46] while China will probably surpass the US economy before 2030.[47] A variety of factors are at play in the Rising Economies, but population growth is key – rising seven times faster than in developed countries.[48] The leading megacities in terms of size will be Tokyo, Mumbai, Shanghai, Beijing, and Delhi and 6 of the top 10 cities will be in Asia.

#### Empowered consumers

The UN and the OECD define middle class in terms of assets, as someone who earns or spends $10–$100 per day and with enough disposable income to buy consumer goods.[45] While we are familiar with the way in which the 19th-century Industrial Revolution transformed agrarian economies, this is a transition on a global scale. Currently, China has 55 million middle-class households, and by 2025 this is expected to reach nearly 280 million.[49] The biggest African economy in 2030 will be South Africa, with a GDP of $1 trillion, only a third of the size of Mexico's, predicted to be the leading Next 11 economy by 2030.[48]

#### Glocalization and politics

The future global political agenda will be defined by glocalization, where the focus will be on supporting growing regional trade, emerging city-states, online communities of choice, and the next generation of entrepreneurs. By 2030, we will see a marked shift towards network politics and coalitions. This diffusing of power among nations will inspire new structures and more local economic models – or "locanomics." Underpinning these hubs will be new business networks and governance models, leveraging global assets in order to capitalize on regional strengths and deliver economic value in a local context.[49]

### Who: The Asian dream

Glocalization is shaping local economies, with the "Asian dream" being a large-scale co-creation process that involves innovation drivers, green practitioners, and cultural experts to challenge the American Dream as a global aspirational growth model. On a more local level, The China Dream project is a vision to reimagine prosperity and deliver a sustainable lifestyle for the emergent middle class in China.[50]

### Where: Prospective economic hubs

**BRIC:** The middle class in BRIC (Brazil, Russia, India, China) nations will grow by 150% to 2 billion people by 2025. China's middle class is likely to represent 75% of the country's population by around 2025. In India, 57% of the population (850 million people) will belong to the middle class by 2030. Brazil's middle class will rise to 127 million people, while Russia's will decline from 71% to 45% in 2030, as people move to a higher income bracket.[48]

**The Next 11:** The rise of the Next 11 nations (Bangladesh, Egypt, Indonesia, Iran, Mexico, Nigeria, Pakistan, Philippines, Turkey, South Korea, and Vietnam) has been described as the "great doubling." By 2030, the combined middle class of these 11 nations will have grown by 116% to 730 million people.[48]

**MINT:** According to Jim O'Neill, who coined the term BRIC, the MINT (Mexico, Indonesia, Nigeria, and Turkey) economies have potential because of their favorable demographics and recent strong growth.[51]

**NORCs:** The Northern Rim countries, Scandinavia, Greenland, Iceland, Russia, Canada, and Alaska, have been tipped to become climate migration magnets. By 2050, increased oil production could make Canada the world's second biggest economy and turn the region into a new powerhouse.[52]

## WHY: RISING ECONOMIES BUSINESS CASE

### People: Increased prosperity

**Better lives:** Half the world's population will have joined the so-called "consuming classes" by 2025,[53] meaning better lifestyle opportunities in everything from education to jobs and health.

**Political confidence:** Middle-class voters tend to have the confidence to engage in civic life, seeking to reduce corruption and improve education and the local environment because they feel they have a "stake" in their community's prosperity.[54]

### Planet: A balancing act

**Hyper-urbanization:** With Asia accounting for 55% of new urban residential construction to 2025, there will be pioneering initiatives on green megacities.[53]

**Sustainable development:** Green construction is a strong driver in the new economies – China's aim to construct over one billion square metres of green buildings by 2015 would account for 20% of all new buildings in its cities.[33]

### Purpose: Cultural sharing

**Storytelling:** Cultural capital becomes increasingly embedded in local thinking strategies. The benefit is a deeper knowledge and understanding of local cultures and customs.

**Soft power:** China's soft power initiatives, such as combining training future African leaders with economic investment in the region, are likely to spark further development and partnership opportunities on a global scale.

### Profit: Tapping new markets

**Affluence:** A huge wave of increasingly affluent population will constitute China's urban majority by 2020.[55]

**Luxury goods:** Asia may account for half the luxury goods market within a decade.[56] However, a joint Indian–US study suggests localized approaches to marketing will be important factors to future success.[57]

## HOW TO IMPLEMENT: Scientific Dimension trends – action points

### 1. Radical Openness

- Recognize that authentic organizations build bridges, not walls
- Have clearly aligned values and business goals

- Grow trust capital through openness and consistency
- Use established platforms to develop transparency and accountability
- Reputation management requires sound ethical foundations
- Continually strive for value-creating solutions

2. The Global Brain

- Make responsive open dialogue central to business practice
- Employ real-time data networks to assist people – be helpful not intrusive
- Connect via the Global Brain, but be transparent about the data you hold
- Invest in meaningful analytics to foster business growth
- Combine data with people, processes, and things to harness opportunities
- Use the power of big data to tackle risk and challenges

3. Green Growth

- Make renewables integral to your strategy and partner with nature
- Connect to people with an inspiring sustainability narrative
- Implement the Circular Economy – reduce, reuse, recycle, rethink
- Design your green road map to drive sustainable systems thinking
- Support the creation of local green jobs and invest in cleantech

4. Rising Economies

- Champion decentralized power and extend your "glocal" networks
- Target the new middle classes with market-appropriate offers
- Add shared value in a local context via sustainable development
- Strengthen your brand storytelling by adapting to local market needs
- Look for new opportunities developing around hyper-urbanization

---

## Trends in the social dimension 2030+

## 5. Smart Living

### What: Intelligent spaces and advanced system thinking

As physical and virtual borders dissolve, seamless transitions and self-defined boundaries in all areas of life will abound. Innovations that combine robotic technology, architecture, and design are the absolute enabler of this trend. Remote working, socializing, shopping, and "on-the-go" monitoring of our home, energy consumption, and health are all enabled by connected technology. Smart living will drive a more sustainable future, inviting people and business to collaborate – with the power balance tipping towards people.

### Self-quantifiable tech

By 2030, a "born mobile" generation, constantly connected to work and home life, will be using wearable or even implantable devices. Wearables already show great promise, especially self-quantifiable and digital cross-analysis tools for health monitoring – with wrist devices as favorites. So far, the device that has gained the most publicity is Google Glass, a hands-free wearable computer. But it is the large-scale projects that will reshape business sectors in years to come. The trajectory of user interfaces is already being plotted and one likely scenario is a hybrid of the invisible and the tactile to offer truly responsive and usable tech that becomes second nature to use.

### Connected domestic helpers

Machines are already having silent conversations in the background of our lives – saving society and businesses money, time, and resources, while enhancing personal control, experience, and wellbeing. By 2030, real-time homes with "connected robotic consumer devices" – for instance learning robots, augmented bathrooms, smart kitchens, and emissions monitoring – will be a commercially viable reality on a mass scale. Technology will be subtle and seamless compared to today, offering helpful suggestions on how to optimize and balance our lives. The immediacy of invisible technology has the potential to change human behavior for the better.

## Who: Enhancing personal control

Consumer devices have huge potential to save time and money. The most promising products to emerge so far – including the thermostat NEST – add value and save money. The more ambitious Samsung Smart Home app promises a means to manage all home technology through one platform. It is already being rolled out through Samsung devices and future apps may support areas as diverse as healthcare products and the humble front door lock.[58]

## Where: Smart living across the world

**Korea:** Smart living environments – known as "silver towns" – include embedded ICT to support seniors at home while reducing the risks associated with independent living.[59]

**Sweden:** The Stockholm Royal Seaside urban regeneration project is a test-bed for new ICTs designed to improve quality of life, grow the local economy, and help Stockholm remain a green leader.[60]

**UK:** London's Royal Docks emerged from the 2012 Olympics as a regenerated, sustainable commercial and residential area and home to Siemens' Crystal building, which showcases technology for the smart city.

**Hong Kong:** The region has become a leader in RFID (radio frequency identification) technology, with smart cards used for everything from transport and building access to library cards and shopping.[61]

**Germany:** Berlin is working with the Swedish power company Vattenfall, BMW, and others to develop vehicle-to-grid technologies that could create a virtual power plant from electric vehicles.[61]

## WHY: SMART LIVING BUSINESS CASE

### People: Human touch

Instinctive interface: Digital technology will soon replicate our bodies, with new evolved instinctive interfaces, giving people personal control of their own health and wellbeing.

Nanotechnology growth: One of the fastest growing sectors, nanotech will impact many industries, including automotive and health, to provide more compact and powerful components.

### Planet: Living networks

Smart grids: By 2020, cities will have invested $108 billion in smart meters and grids, energy-efficient buildings, and sophisticated data analytics to connect homes, cars, and smartphones with power suppliers.[62]

**Mobile technology platforms:** By 2050, cars will be able to act as data platforms linking services such as healthcare, security, and energy consumption. By 2040, self-driving cars minimizing input from the driver could be a reality.[63]

### Purpose: Positive behavior and safety

**Embedded values:** New infrastructures can embed specific values to support and promote sustainable positive behaviors through information feedback loops, behavior modification, and "gamification" strategies.[64]

**Next-gen robots:** While we expect them in sci-fi films and on the factory floor, robotics will become part of our daily lives, helping to reduce risk and assist with day-to-day tasks.[65]

### Profit: Analytics opportunities

**Smart clothing:** Sales of wearable tech jumped almost 300% in 2012, and we bought 8.3 million fitness trackers, smart watches, and smart glasses. By 2018, global sales of smart wearable devices will exceed 480 million.[66]

**Auto analytics:** Apple is already disrupting self-sensing in the same way it disrupted music – with an entire digital monitoring platform for medicine, fitness, and wellness.[67]

## 6. Global Citizens

### What: Home is anywhere

By 2030, global migration could result in workforce and talent shortages in emerging and developed countries. This is why Global Citizens are so important to future business because, rather than being tied to one locality, this constantly evolving group of well-informed and well-connected Millennials see a world without boundaries. Global Citizens are on a mission to build meaningful and self-empowered lives and seek out career and lifelong learning opportunities. Richness of experience is a key motivational driver in their technologically enabled reality, but also a facilitator of exchange between cultures that allows them to embrace diversity and flexibility. For Global Citizens, the "job for life" is a redundant concept as they seek out opportunities that develop their careers and aid a mobile lifestyle.

### Global hyper-mobility

Alongside traditional factors, such as jobs and wage levels, future migration will be fueled by the desire for personal development and social proficiency – enabled by smart technology and hyper-mobility. The Global Citizens' vast networks support an open WE mindset, in which like-minded individuals collaborate to achieve common goals. For them, independence is achieved by forging new bonds within the working and social environments they operate in so seamlessly. It is predicted that there will be an increase of 50% in international assignments by 2020,[68] and this group will use the opportunity to boost career and life experiences.

### Equality and profit growth

This rising demographic of Millennials will account for 46% of the American workforce by 2020[69] and it is essential to engage with this group because of the influence they wield. With many more female influencers shaping politics, business, and society, we see the rise of fresh approaches to trading, collaborating, and creating value – making the female factor central to a more inclusive and prosperous future. Currently, only 30% of European entrepreneurs are female,[70] but the future will be more gender balanced, with two-thirds of university graduates in developed economies being women by 2020.[71] Encouraging equality will bring transparency, flexibility, and fluidity to the business world, alongside better bottom lines.

## Who: Nurturing talent networks

**"Go local":** A study of nine cities by Dell concluded that the key to maximizing human capital and extended entrepreneurial ecosystems is quality networking opportunities – online and face to face.[72]

**"Project Shakti":** Unilever in Hindustan cultivates female entrepreneurship to ensure women are change agents and role models within their communities – a side benefit is growing market share for the company.[73]

## Where: Global talent shortage

**China, Western Europe, and the US** are key economic regions, but a serious shortage of qualified employees means the demand for talent will intensify up to 2030.[48]

**Nordic countries** are forecast to remain relatively unaffected,[38] compared to Asia, in the face of climate change, and may be well placed to attract mobile youth and talent to invigorate an aging population.

**Iceland** emerged from a seismic economic crisis with a culture of fresh values driving change. Women are leading this and their high participation in politics and the workplace is seen as a key reason why recovery has exceeded expectations.

## WHY: GLOBAL CITIZENS BUSINESS CASE

### People: Skill strategies

Skills blueprint: China's 10-year plan is a blueprint for developing a highly skilled workforce.[74]

Workforce shortage: According to the World Economic Forum, an estimated 10 million jobs within manufacturing organizations cannot be filled today due to a growing skills gap.[75]

### Planet: Value-driven youth

Global talent: 50% of Millennials surveyed want to work for a business with ethical practices.[76]

21st-century working: Remote working saves resources, enables the work–life balance, and widens the talent pool. Major companies, including Xerox, Dell, and Amex, are increasing remote job opportunities.[77]

### Purpose: Meaningful opportunities

Business with purpose: Improving society is the most important purpose of business, according to 36% of Millennials.[78]

Longer working: We are witnessing how the current 50+ segment – or the "No-age" Global Citizens – are working longer and seeking new opportunities through traveling.

### Profit: New balance

Global income increase: The World Bank estimates that a 3% increase in the stock of migrants by 2025 would grow global income by 0.6%, or $368 billion.

The right mix: A gender-balanced workplace composition can achieve operating margins twice as high as those in less diverse organizations.[79]

## 7. Betapreneurship

### What: Not perfect is good enough

The world is currently seeing a wave of new business structures, with disruptive innovation and social entrepreneurship becoming the norm. Betapreneurship means operating in a trial-and-error process, where creativity and improvisation fuel problem solving and ideation. This approach is a way to foster innovation culture and job creation, and it's a philosophy that will inform start-ups and larger organizations in the future. Over two-thirds of Millennials see themselves working outside traditional organizations at some point in their career.[76] They are already playing a key role in transforming today's entrepreneurial culture into Betapreneurship and a more community-driven proposition that will improve social mobility and local growth.

#### Free radicals shape the future

People who thrive on flexibility and feel most productive when they are fully engaged are more likely to work for progressive organizations or create their own start-ups. These 21st-century "free radicals" are Betapreneurs, who view failures and setbacks as a learning opportunity.[80] Self-reliant and extremely potent professionals, they will profoundly influence future business, whether they work solo, in small teams or within large companies. They demand freedom to experiment and add creative input that moves ideas forward, which is why, according to Deloitte, almost four out of five Millennials are strongly influenced by an organization's approach to innovation when deciding on a future employer.[76]

#### Cultivating "intrapreneurial" culture

Millennials see the biggest barriers to innovation as management attitude (63%), operational structures and procedures (61%), and staff skill sets, attitudes, and diversity (39%).[76] It is essential for organizations to present themselves as attractive to this group, which is why innovation-directed businesses hold product and service innovation contests, invest in start-ups, and bring on entrepreneurs in residence – also known as "intrapreneurs."[81] Already, intrapreneurship is fueling innovation in many of the world's most successful organizations, and any smart 21st-century business will cultivate this culture to develop effective leadership and succession strategies, grow, and diversify in order to gain market advantage.[82]

## Who: Encouraging more start-ups

The EU's *Entrepreneurship 2020 Action Plan* is designed to encourage more citizens – particularly young people – to develop start-ups. The idea is to give recognition to entrepreneurs so they contribute to Europe's economy. Cultural factors may inhibit people's entrepreneurial spirit, which needs to be developed from school education on – by encouraging experimentation, even when it fails. Red tape is another major hurdle, and sources of funding and investment need to be made readily available. This may include exploring alternative business models, such as the cooperative enterprise.[83]

## Where: Inspiring start-up ecosystems

**Bangalore:** India's most successful IT hub since the 1970s provides a fantastic entrepreneurial environment. It is thriving because of its existing business clusters, economic opportunities, and tech-savvy, educated population.[84]

**Tel Aviv:** Ranked second in the world by the Startup Ecosystem Report, just after Silicon Valley on the list of best places to establish a business.[85]

**London:** The world's seventh most influential start-up ecosystem.[85] As of 2012, Silicon Roundabout (Tech City) in East London had more than 3,000 tech firms, employing up to 50,000 people.[86]

**São Paulo:** The Latin American hub is now ranked 13th in the world for its start-up environment. But it has competition from Santiago (20th), supported by the government-backed Start-up Chile program.[87]

**Boulder, Colorado:** Has more technology start-ups per square mile than any other US city, according to Brad Feld, who also argues that every city of 100,000 plus needs a start-up community in order to be healthy.[87]

### WHY: BETAPRENEUR BUSINESS CASE

#### People: Thriving on diversity

Multiculturalism: Almost two-thirds of Silicon Valley professionals working in science and engineering were born outside the US, more than twice the national average for similar professions and education levels.[88]

**Critical growth factor:** Finding and keeping the right talent is the most critical factor to business growth, according to 97% of CEOs[68] – one good reason why nurturing intrapreneurship makes sound economic sense.

### Planet: Encouraging start-ups

**More know-how:** Many graduates with valuable specialist knowledge want to start a business, but lack the know-how of balance sheets and business plans. Universities are responding with postgraduate qualifications in entrepreneurship.[89]

**Nurturing future inventors:** The Imagination Foundation works to develop children's inventive skills, helping them to create the future they dream of – and it started with a small boy's cardboard arcade.[90]

### Purpose: Creative leadership

**Dynamic growth:** Intrapreneurship is the key to dynamic growth and change and, for Millennials, it's an opportunity to develop their leadership skills while inspiring positive change.[82]

**Social intrapreneur typology:** Successful intrapreneurs are motivated by the incubation and delivery of new ideas that add value to society and the bottom line.

### Profit: Innovative thinking

**Made to last:** "Antifragile" businesses adapt to challenges and use them positively, whereas resilient businesses resist shocks, but don't adapt, making them more vulnerable.[91]

**Intrapreneurial results:** By 2009, half of all Google's products originated from the "20 percent" program – where top managers work on self-initiated innovation and problem-solving projects one day a week.[81]

## 8. The No-age Society

### What: Changing mindsets = positive aging

The UN forecasts that 2 billion people will be 60+ by 2050, that's 22% of the world population. So, in effect, 10,000 baby boomers will join the seniors category daily for the next two decades.[92] "No-age" lifestyle patterns are set to become a reality with the arrival of a 5G (five generations working together) workforce, hence a more positive mindset towards aging will be needed – one in which seniors are viewed as a resource not a burden. A vast army of fit seniors will not only contribute to the economy and a healthier welfare system, but become living advertisements for the benefits of active and balanced lifestyles when it comes to maintaining financial independence and aging well.

### Health for everyone

Chronic diseases will account for almost 75% of all deaths worldwide by 2020, according to WHO,[93] but mHealth (mobile health) intelligence is evolving as an accessible resource for all and a strong business opportunity. Personal digital diagnostics for balanced health, fitness, and diet deliver tailored solutions to help us lead better lives, and healthcare professionals will become involved in designing monitoring solutions for prevention and awareness, rather than just healing. Advanced systems thinking brings together diverse stakeholder groups in the quest to develop technologies that are adaptive, collaborative, and resilient enough to inform, educate and, ultimately, improve our wellbeing throughout our lifetime. An added benefit comes in reduced costs and pressure on welfare systems.

### The social impact of longer lives

By 2050, one in every five older person will be aged 80+.[94] A baby born in 2011 is almost eight times more likely to reach 100 than one born in 1931.[95] The generation beyond that could live even longer, and the median age will rise to 34 years worldwide and 44 years in developed economies by 2030.[48] One of the key society-wide impacts will be age blurring in terms of cultural outlook and lifestyle choices. A potential downside will be that, with vast amounts of GDP being diverted to reverse aging, we might experience a potential growing divide between the younger-to-middle cohort and senior citizens in some regions of the world.

## Who: A future geared towards seniors

**The rise of "olderpreneurs"**: In the UK, around one in five people aged over 50+ is self-employed and the number of first time businesses started by this group is rising – a trend also becoming more visible in the US. They have been termed "olderpreneurs," and they set out to counteract rising retirement ages, shrinking pension funds, and lack of employment opportunities by going it alone. They are likely to inspire others, especially since senior start-ups appear to have a high success rate. There are many reasons to choose self-employment at any age, but with more funding and support networks provided by governments and NGOs, this sector of self-reliant and savvy entrepreneurs will grow in strength.

**No-age design:** In Japan, home to a rapidly aging population, Toyota is working with Professor Kawashima, who developed brain-training games for Nintendo, to create intelligent cars. The idea is that the car monitors brain activity in the senior driver to learn the user's driving patterns and counteract unusual or dangerous activity. Similarly, cycle equipment maker Shimano is developing bicycles with handles and gearing designed for arthritic hands, and Panasonic has launched a robotic bed that can transform into a wheelchair to enable people to get up without assistance.[96]

## Where: Senior nations around the world

**Japan:** 23.3% of the population were aged over 65 in early 2011 and life expectancy at birth was 83.1 years in 2010, the highest in the world. The other top 10 "oldest" countries as of January 2011 are in Europe, with Monaco in second place, followed by Germany, Italy, Greece, and Sweden.[97]

**Northern Europe:** Good places to look for inspiration on the No-age society are Sweden (first place) followed by Norway, Germany, the Netherlands, and Canada. According the Global AgeWatch Index, these are the best countries to grow old in – based on four markers: income security, health status, education and employment, and enabling environment. At the bottom of the list are Pakistan, Tanzania, and Afghanistan.[98]

**UK:** An emerging trend of career reinvention in retirement means that almost 1 in 10 (8%) of retirees had chosen to change careers after official retirement and 1 in 20 had started their own business.[99]

## WHY: NO-AGE SOCIETY BUSINESS CASE

### People: Longevity and made to measure

**E-health apps:** The use of medical services will change dramatically as electronic health records proliferate.

**Tailored for women:** Women are likely to pursue work for longer than men, and increasingly they will look for tailor-made products and services in all areas of life – from hotels to work technology.

### Planet: Considered consumption

**Built to last:** Seniors' attitude to resources (wealth and planetary) will make them sophisticated and caring consumers who look to buy reasonably priced quality goods.

**Gray behavior:** Longevity means gray consumers will still be investing, perhaps planning a green, tech-enabled car purchase at 70+ and a smart inclusive home adapted to fit their specific needs in their sixties or beyond.

### Purpose: Staying active longer

**"SYLO" generation:** Having spent their adult lives in a world of mobility and connectivity, the "staying younger longer" (SYLO) generation won't settle for traditional retirement and will seek out opportunities to maintain their mental and physical agility.[100]

**No-age seniors continue working:** Those who don't embark on a new career path might choose to "give back" through community and civic life.

### Profit: No-age business opportunities

**Financial control and happiness:** A survey by Aviva found that health is considered twice (85%) as important as money (42%), and "financial control" – rather than wealth per se – is the key to happiness.[101]

**Investing in gray business:** Of business enterprises started by those in their fifties, over 70% survive for at least five years, compared to only 28% of those started by younger people.[102]

## HOW TO IMPLEMENT: Social Dimension trends – action points

### 5. Smart Living

- Collaborate to help people benefit from "smart living" technology
- Rethink "human" interfaces, build in improved personal control
- Assist positive behaviors by creating tools to optimize lives
- Enable remote work and monitoring via connected tech solution
- Promote balanced living – engage in smart systems thinking

### 6. Global Citizens

- Empathic leadership is required to attract the "born mobile" generation
- Flexible work options maintain a vibrant workforce, including Millennials

- Nurture entrepreneurship and support an open WE mindset
- Boost career opportunities by offering personal development
- Create an inclusive workplace – inspire equality and female leadership

7. Betapreneurship
- Creative leaders encourage new disruptive business models
- Promote social entrepreneurship to nurture mobility and local growth
- Develop trial-and-error learning processes for true innovation culture
- Give people freedom to experiment as intrapreneurs
- Employ and develop diverse talent to build long-term growth

8. No-age Society
- Champion a 5G workplace – promote active and healthy lives
- Welcome seniors as a valued resource with flexible work solutions
- Encourage No-age community activities for an inclusive culture
- Support lifelong learning in the workplace and communities you serve
- Champion digital health diagnostics and smart work/life tools

---

## Trends in the emotional dimension 2030+

## 9. Better World

### What: Conscious consumption

In order to create an economic landscape that comes closer to matching people's future "good life" ideals, we need pioneering new economic models. Citizen awareness and civic power mean people now expect leaders in government and business to evidence a proactive stance, together with accountability in relation to workforce, suppliers, waste reduction, and responsible management of natural resources. Currently, this is being discussed across the world and is set to inspire a reappraisal of present performance parameters. We need to look beyond a bottom line based solely on profit and market share and nurture a better glocal economic system, underpinned by a 4P foundation that serves the ecosystem in which it operates – one where the impact on people, planet, and purpose are judged alongside profit.

### Driving a betterness agenda

Betterness isn't a utopian perspective, but rather one where people and organizations become active participants in creating a positive future. A strategy that demonstrates values-based thinking will become essential to all business models in the future. Fostering a sense of community around business develops live and flourishing networks, which is why forward-thinking organizations now invite collaboration to promote inclusive values. More than ever, we are aware that we cannot rely on government alone when it comes to the future of our environment, welfare system, and quality of life. Businesses have a great opportunity to lead change – empowering people and the local communities they serve to become prosperous positive influencers as they make better and more sustainable choices.

### Rise of sharing and fix-it culture

A culture of bartering, sharing, and fixing things – rather than buying things – is thought to create value by fostering a more sustainable and fair society. This global movement favors access to goods and services – also known as "collaborative consumption" – over ownership. The trend is set to influence every area of tomorrow's society and challenge current business models. Connective technology has enabled the sharing of idle and surplus resources, reconnecting people and communities while potentially reducing their outgoings. Fundamentally, people are realizing that sharing and fixing things makes us feel more content, is more sustainable, and supports our idea of what constitutes a better world.

## Who: Responsible organizations

**Less is more:** Becoming part of the solution, not the problem, is why Patagonia and eBay partnered for the Common Threads microsite that enables people to sell or buy second-hand Patagonia clothing and champions the reduce, reuse, recycle, and repair message.

**Plan A:** UK retail giant Marks & Spencer launched its Plan A ("because there is no plan B") strategy back in 2007. The company's aim is to become the world's most sustainable retailer, and this strategy has disrupted retail as usual. Plan A set out to achieve 180 sustainability commitments by 2015.

**Pioneering repair culture**: iFixit is a growing global community of people helping each other to mend things. Offering free repair manuals and advice, its slogan is: "Let's fix the world, one device at a time." Based around the manifesto that repair is better and more sustainable than recycling, it is urging consumer electronics manufacturers to "set devices free" by unlocking them in order for people to do repairs themselves.

### Where: Thriving sharing economies

**The Netherlands** is developing a thriving sharing economy, with 84% of the population sharing. The most common swapped goods and services are food, magazines, and knowledge and the biggest motive to share is, by far, to help others.[103]

**Berlin** is the largest one-way car-sharing city in the world. A flexible business model enables users to pick up the car nearest to them and leave it at a convenient location.[104] Car sharing reduces pressure on overcrowded infrastructures, and schemes have been launched in, among others, São Paulo and Beijing (2009), Hangzhou and Istanbul (2011), Mexico City (2012), and Bangalore (2013).[105]

**Seoul** is the world's first official sharing economy. With 60% of residents owning a smartphone, collaborative culture is improving life in this overcrowded city. Initiatives include car, tool and luggage sharing, spare room rental, and citizen access to idle public resources.[106]

## WHY: BETTER WORLD BUSINESS CASE

### People: New citizenship

Leadership challenge: Almost half of Millennials feel governments are having a negative impact on areas identified as among the top society challenges, including resource scarcity and unemployment.[78]

Trust 2.0: According to US community-share membership organization Peers, over 75% of people who share online say they have greater trust in their community.[107]

### Planet: Bottom-up empowerment

Real costs: It has been predicted that, by 2050, the price of a product in Denmark will include the environmental costs associated with production.[38]

**Collaborative movement:** Resource sharing is on the increase; by 2012, *The Ecologist* estimated that there were 161 resource-sharing partners globally, with a reach of over 60 million people in 147 countries.[108]

### Purpose: Sharing is caring

**4P support:** A vast majority (87%) of global consumers believe business should place at least equal emphasis on social interests as business interests; "purpose" has increased as a purchase trigger by 26% since 2008.[109]

**Empowerment:** Collaborative consumption is essential to sustainability, productivity, innovation, and competitive advantage, but a huge side benefit is the empowerment of localities, social groups, and individuals.

### Profit: Collaborative consumption

**Business model innovation:** Disruptive models driven by Betapreneurs are providing platforms to increase performance and sustainability – they see collaboration, sharing, and repairing as a way to optimize resources and minimize waste.

**New market opportunities:** House-sharing platform Airbnb claims that it added $632 million to New York City's economy in 2013.[110]

## 10. Lifelong Learning

### What: Curiosity drives talent

One of the core drivers shaping the future will be lifelong learning, as demographic changes and a new generation of mobile talent deliver a No-age society with a diverse 5G workforce. A US study concluded that 65% of primary (grade) school students will work in jobs that "haven't been invented yet."[111] This suggests that new approaches to education are critical; lifelong learning has the capacity to democratize education globally and improve competitiveness, driving social mobility and entrepreneurship. Penetration of the Internet, smartphones, and tablets is already bringing disruptive education models to expand the traditional classroom or lecture hall into a wider panorama of virtual interactive learning environments, where educators lecture less and mentor more.

#### "Digital-centric" learning

The Global Brain will drive a lifelong learning agenda through digital knowledge sharing. "Campus-centric" old schools will evolve to be where people are and that means embracing "digital-centric" new schools. The proliferation of Massive Open Online Courses (MOOCs) has already made learning at a higher level available to all – a transformative ICT platform that has great value as it provides a low-cost forum for the dissemination of knowledge, discussion, and professional development skills. Digital models are also heralding an unprecedented democracy of learning that is set to reshape education in the future and make knowledge development more inclusive – reaching across demographic and income brackets of society into even remote communities and cultures.

### New educational benchmarks

Looking at tomorrow's workplace, we are moving away from "fixed" and into more "flexible" mobile environments, where the technology-driven knowledge economy will place a premium on very different attributes – hands-on, cross-disciplinary, and continually updated skills acquired through self-learning. Keeping pace is essential for competitiveness; by 2020, 35% of jobs in the EU are likely to require a higher education qualification,[112] and demand for highly qualified people is projected to rise by almost 16 million to 2020.[112] What we learn is also up for review, with traditional education passports – including the MBA – being challenged as Asian nations develop lower cost models with a practical emphasis on local market conditions.[113] Meanwhile, developing nations are pioneering lean models of university education with a reduced cost structure that could eventually influence current Western models.[114]

## Who: Equipping future leaders

Singularity University, an unaccredited institution located in Silicon Valley and backed by Google, Cisco, and Nokia, among others, is led by Peter Diamandis and Ray Kurzweil and works with entrepreneurs, technologists, and leaders worldwide. Its mission is to educate, inspire, and empower current and future leaders by preparing them to address extraordinary global challenges to come. Its executive program was developed to give people the tools to predict and evaluate how disruptive innovation will transform industries, companies, careers, and lives.[115]

## Where: Financial literacy for all

The current emphasis is on business-focused courses, with over 40% of MOOCs being in the subject areas of business, economics, and social sciences. Little wonder, then, that organizations such as the IMF see the potential of democratizing financial literacy. It is launching economics courses in member countries targeting civil servants and government officials, and will roll these out to the general public to improve knowledge of financial issues.[116]

### WHY: LIFELONG LEARNING BUSINESS CASE

#### People: New structures

**Supporting longer working lives:** Participation rates of older workers in most OECD countries were higher in 2008 than in the 1970s,[117] with later retirement suggesting that people will need ongoing skills development and training to support their employability.

**Made-to-measure learning:** In the future, students won't be forced to adapt to a "one-size-fits-all" model; technology will allow the system to adjust to every student's learning and career needs independently.[118]

#### Planet: The global classroom

**Multilayered models:** Augmented reality and geocoded information will be available any place any time – inspiring social structured learning models.

**User-centric approach:** Organizations will have to move from a device and channel-centered to a user-centric approach, in which they adapt their messages to multiple devices.

#### Purpose: Social engineering

**Knowledge + social value:** Future MBA programmes will see sociocultural exchange schemes and preventive holistic health included in the curriculum.

**Diplomacy education:** MOOC Camp, an initiative by the US Department of State, invites MOOC students to gather at its local embassies and other spaces to participate in discussions led by, among others, Fulbright Scholars.[119]

#### Profit: Education ecosystems

**Enabling learners:** A new education ecosystem requires a broad range of enablers, including IT, network, content, hosting and data management services.

**Education value:** Mobile learning is expected to be a $70 billion market by 2020 and will enhance learning outcomes worldwide.[120]

## 11. Social Capital

### What: We are in this together

The World Bank[121] defined the value of social capital as "not just the sum of the institutions which underpin a society – it is the glue that holds them together." Using this definition, it is clear that by cultivating cultural capital, we enable the social capital growth that is essential for a thriving business. This trend marks the growing recognition that networks between individuals or groups are valuable in social and economic terms. Emerging studies of what makes societies function question the accepted wisdom that measures prosperity in purely fiscal terms. In tandem, growing social entrepreneurship and wellbeing economics are informing the next generation of business leaders. This challenges business models to include strategies contributing to a more equitable world. Increasingly, people want to know and "feel" exactly why they should engage with your organization; hence, the capital P in leadership will be purpose.

#### Collaborative communities

Growth in participatory culture – as seen in crowd sourcing, community participation, affinity networks, and volunteering – is the positive measure of people's desire for involvement and their aspiration to "give back" and "be better together." In its broadest sense, social capital encompasses the links, shared values, and understandings in society or within organizations that enables individuals or groups to establish trust and work together in meaningful ways. Organizations tap into a wealth of opportunities by developing social capital that revolves around autonomy; in practice, this means inviting a culture where affinity and vision inspire people to contribute for collective gain.

#### Culture of purpose

A purpose-driven organization is one that allows people to contribute with their thoughts and ideas, fostering an open, collaborative working environment where everyone feels recognized and respected for their role and talents. This means creating meaning and value beyond the bottom line, but it also implies that it is imperative for organizations to understand the social and cultural context in which they operate. Most crucially, they must recognize that society – and in particular our cultural environment –

is increasingly made up of live social networks. These networks continually scrutinize organizations by determining their worth and performance in terms of values, ethics, and social engagement.

## Who: Inclusive welfare systems

There are indications that the Nordic region is doing well in growing and retaining talent, with Scandinavia occupying four of the five top places in *The Global Talent Index Report: Outlook to 2015*,[122] alongside the US; both regions are cited in particular for their education provision. Social capital underpins the strong welfare system in the whole Nordic region, and research from the National Research Centre for the Working Environment[123] in Denmark indicates a relationship between corporate social capital and economic stability and performance over the long term.

## Where: The Social Progress Index

Harvard professor Michael Porter, also known for the "shared value" concept, created the Social Progress Index (SPI) to support corporate and government thinking. Initiatives cited in these rankings include access to information and communications policies. The three dimensions explored are basic human needs, foundations of wellbeing, and opportunities. New Zealand tops the 2014 SPI, although if individual metrics are explored, the highest ranking country for improving "foundations of wellbeing" is Switzerland, followed by Iceland, the Netherlands, and Norway. The UK is ranked 19th and the US is 36th. China is placed 90th and India 102nd of 132 countries, suggesting that a key task for Rising Economies in years to come will be to improve human wellbeing and corporate and government accountability through better access to information.[124]

## WHY: SOCIAL CAPITAL BUSINESS CASE

### People: Together we are stronger

**The power of networks:** In a patchwork society, it is people's networks that enable them to belong. These networks continue to grow, making a strong online presence and real-time responsiveness essential factors for the success of all 21st-century organizations.

**Corporate social innovation:** Projects grounded in community and locality – led by stakeholders not organizations – will become an essential part of any CSR agenda.

### Planet: Collaborative ecosystems

**Caring nations:** Responsibility for our planet will be viewed as a truly shared concern, as countries join economic, political, and social collaborations by 2030.[48]

**Cross-border learning:** Sharing tools and information will bring cross-border opportunities to build stronger support ecosystems locally and disseminate examples of best practice.

### Purpose: Tuning into people's values

**The "why" in purpose:** The most successful organizations will explain their purpose beyond profit, conveying to stakeholders through their mission and strategy their real value to society.

**Attracting "thought leaders":** With a continued global talent shortage, organizations will look to attract and retain "thought leaders" by developing a mission that tunes into people's beliefs and cultures.

### Profit: The rise of the social enterprise

**Principles for profit:** Many new-gen entrepreneurs, including co-CEO and co-founder of Whole Foods Market John Mackey, argue that data proves that conscious businesses will outperform traditionally run companies by a wide margin in the long run, also creating value on multiple levels.[125]

**Strategic alliances:** Partnerships between the voluntary and private sector are emerging as a core trend, and are set to become even more widespread as they are recognized as a key business opportunity over the coming decades.

## HOW TO IMPLEMENT: Emotional Dimension trends – action points

### 9. Better World

- Pioneer new business models to lead change
- Use values-based thinking that considers people, planet, and purpose
- Drive a "betterness" agenda and act as a role model
- Sharing is caring – participate in the sharing economy
- Collaborate to stimulate inclusive values and flourishing networks

### 10. Lifelong Learning

- Cultivate a 5G workforce and enable continual skills upgrades
- Embrace digital-centric education models to democratize learning

- Promote social values by developing cross-disciplinary skills
- Give people the right transformation tools to inspire and empower them
- Great leaders subscribe to a lifelong learning agenda

11. Social Capital

- Be an empathic organization by promoting strong values and ethics
- Grow social capital – inspire people to contribute for collective gain
- Be an engagement brand by nurturing community and cultural capital
- Harness the value of the "social enterprise" to deliver "real" progress
- Foster a participatory culture and be a change-maker organization

## Trends in the spiritual dimension 2030+

## 12. The "Good Life"

### What: Global wellbeing strategies

Inextricably linked to all the key trends, the "Good Life" is certainly a key theme for the 21st century. However, the core challenge in achieving the Good Life is: How do we balance natural, human, social, and economic capital to sustain the wellbeing of a society over time? Since 2006, Gallup has polled more than 160 countries, categorizing them as "thriving," "struggling" or "suffering." Measures include the Job Climate Index, shown to correlate with community wellbeing, entrepreneurism, and economic energy. In 2011, the United Nations General Assembly called on all its member states to strive to achieve not just economic development but also to factor in happiness and wellbeing in all public policies. The resolution stated:[126] "the pursuit of happiness is a fundamental human goal," in line with the globally agreed targets known as the Millennium Development Goals for 2015.

### New prosperity measures

Across the world, institutions, politicians, economists, and businesses alike are rethinking the definition of prosperity. The core philosophy of the Good Life is that what is good for community also benefits businesses, and this principle will inform tomorrow's leaders. Measuring gross national happiness (GNH) alongside conventional gross domestic product (GDP) is a key suggestion – and one put forward by the Bhutanese prime minster Jigmi Thinley in his opening speech at the first UN conference on happiness in spring 2012.[127] Back in 1971, Bhutan rejected the idea of measuring national progress and prosperity through GDP only. Instead, the tiny landlocked country, bordered by the powerhouses of India and China, installed a GNH index based on four pillars: equitable social development, cultural preservation, conservation of the environment, and promotion of good governance.[128]

### "Enoughism" and contentment

A sense of authentic wellbeing can inspire and enhance prosperity in all areas of life – productivity, social connectivity, and improved public health – but more importantly, it motivates people to lead more fulfilled lives. Psychologist Mihaly Csikszentmihalyi describes "flow" as the secret to happiness, saying it is "an 'ecstatic' state of mind where we are completely absorbed and fulfilled by what we do."[129] In a business and work context, this is called "passion" – derived from doing our best and having a clear sense of purpose. The world-renowned Zen master Thich Nhat Hanh says:[130] "We need a real awakening, enlightenment, to change our way of thinking and seeing things." He believes that our addiction to consumerism creates an illusion of happiness, causing a lack of meaning and emptiness in people's lives. His model for the Good Life is one where meaning replaces consumption as the ultimate goal.

## Who: Perception shifters

At Harvard Business School, the course entitled "Positive psychology as a catalyst for change" taught by Tal Ben-Shahar, exploring topics such as happiness, self-esteem, empathy, love, meaning, creativity, achievement, and humor, has been so well subscribed it is already inspiring a new generation of business leaders.[131]

## Where: Pioneers of the good life

**Uruguay** has taken a lead with strong initiatives by President José Mujica, renowned for his down-to-earth and liberal leadership style, and was named country of the year in 2013 for its policies to improve quality of life.[132]

**Brazil**, inspired by Bhutan's example, has launched its own version of GNH – *Felicidade Interna Bruta* – with a variety of surveys and activities designed to measure and then improve lives.[133]

**Scandinavia** is one of the happiest regions in the world based on a number of comparative research ratings, including the OECD Life Satisfaction survey.[134] While studies will continue to attempt to understand how it has managed to combine economic and emotional wellbeing at societal level, the Danish proverb "Alt med måde" or the Swedish "Lagom är bäst" – roughly translated as "All things in moderation" – best explain the Nordic approach to the Good Life. What all countries share is less disparity between rich and poor, plus a welfare system that prioritizes the individual in order to cultivate social wellbeing.

## WHY: GOOD LIFE BUSINESS CASE

### People: Boosting happiness levels

Happiness genes: Abraham Lincoln is reported to have said: "Most people are as happy as they make up their minds to be." Research suggests that this is true because happiness or subjective wellbeing is made up of approximately 50% genetic factors, 10% circumstances, and 40% our activities and choices – giving great scope to improve people's happiness levels.[135]

Creativity breeds happiness: A study of college students from the University of North Carolina-Greensboro found that: "Engaging in creative pursuits allows people to explore their identities, form new relationships, cultivate competence, and reflect critically on the world. In turn, the new knowledge, self-insight, and relationships serve as sources of strength and resilience."[136]

### Planet: Balancing GDP with GNH

Nature is our teacher: Bhutan's "green schools" project puts sustainability on the curriculum by teaching its revolutionary "happiness" model to all young people.[128]

Global governance: In the future, happiness will play a greater role in development. Worldwide, there is a rising demand for public policies to be more aligned with people's values and what really matters to us.

### Purpose: Quality of life factors

Mindfulness: Wellbeing coaches now teach organizations how mindfulness can be applied to everything from personal development, leadership strategies, and branding, to creating happiness at work in order to cultivate purpose and increase team performance.

**Wellbeing and aging:** While energy, physical appearance, and mental agility may decline later in life, new theories suggest that happiness actually increases with age.[137]

### Profit: Happynomics

**Happiness as a business strategy:** Happy people are more engaged, innovative, and focused. The productivity of happy people is said to increase by 40–50% in service and creative fields, potentially a significant boost to business revenue and the bottom line.[138]

**Happy employees pass it on:** Happiness is good for the bottom line. A study of service departments found that employees who score highly in life satisfaction are also significantly more likely to receive positive customer feedback ratings.[139]

---

## HOW TO IMPLEMENT: Spiritual Dimension trends – action points

---

### 12. The Good Life

- Recognize that what is good for communities also benefits businesses
- Be a catalyst of positive change – focus on purpose and meaning
- Balance human, social, environmental, and economic capital
- Foster down-to-earth leadership to inspire authentic wellbeing in others
- Motivate people to lead better lives – implement a wellbeing index
- Cultivate happiness – the ultimate currency for a thriving organization

---

Chapter 6

# Practical trend mapping: focusing on people

*You need to truly understand the ecosystem in which you operate, because only then can you stand out and deliver experiences that are meaningful to people.*

The economist Fritz Schumacher observed in his 1970s' global bestseller *Small is Beautiful: A Study of Economics as if People Mattered* that we must look holistically at the world. Long before environmental and political awareness of issues such as global warming and protecting ecosystems had become mainstream, he pioneered the concept of natural capital in a world of finite resources. His ideas proved to be prescient, since the book was published in 1973, the year of the global oil crisis. Schumacher understood that our relationship with the planet was a challenge and believed that helping people to help themselves by making better choices should develop through our inner wisdom, not as a result of intervention from state, science or technology. In Schumacher's ideal vision of the world, the economy was underpinned by small, autonomous businesses, committed to ecology and peace or, as he termed it, "Buddhist economics."[1]

## The people factor

Purpose-driven leadership focuses on having a meaningful influence on people and planet by engaging with the world in a broader context. To do this, organizations need to put people back in the center of their business model – where they belong – alongside a balanced future-focused agenda with long-lasting positive impact. In fact, there is hard evidence that companies operating with a strategy based on "conscious capitalism" – with the greater good of all at its heart – produce better bottom lines. Employees, customers, and investors are now benchmarking companies on a whole new set of criteria. For well over a decade, Kjaer Global have observed people's evolving need for meaningful experiences, and we explored this theme in an interview for the 2010 book *Meta Products*, about products and services connected directly to the Internet of Things (IoT):

> Understand that people – whether they work for you, use your services or buy your products – have higher standards and more complex decision-making processes than ever before. ... so whatever your offer, present it an ethical and meaningful package.[2]

A measure of a 21st-century organization's true "value" is how it engages with people and the world. As author Simon Sinek puts is: "People don't buy what you do; people buy why you do it."[3] Being loved means leading by example. Your actions are reflections of your values and, when you share your beliefs, you will attract people who share them. A 4P bottom line is not just another communication strategy, but a wholehearted mission, where the aim is building Social Capital (people), sustainability (planet), and clear, meaningful goals (purpose), balanced by profit.

For far too long, the mantra "I consume, therefore I am" – encouraged by governments, economists, and businesses – fueled the archetypical dream of "more is better" and drove the economic agenda in the West. We are now on the brink of a total reappraisal of these goals, questioning the logic and values behind the current models based on overconsumption and unsustainable growth. These are grassroots-driven issues that must be addressed by governments and organizations alike to bring about the change that people are expecting.

In his 2011 book *Betterness: Economics for Humans*, Umair Haque describes an important rationale for successful businesses of the future – what he terms the "post-capitalist economy." In this description, he argues that we need to rethink the future of human exchange and our economic paradigm. In short, we need to get out of business and into betterness. He says:[4]

> If you can't demonstrate that at the very least and at the barest minimum, you're not harming people, nature, communities, society, or tomorrow's generations, forget about vanquishing your rivals; you probably won't have a seat at the table.

## New performance parameters: fresh opportunities

The combination of global dynamic forces and evolving political and economic systems is having a massive impact on people's beliefs, value sets, and behavior. Globalization has increased the interconnectivity and integration of the world economy by fueling new economic players such as India and China. But it has also resulted in a far wider range of economic models, where our capacity for global governance has become highly fragmented. This paradox has resulted in a growing divide between nations on how to promote sustainable, inclusive growth. As the World Economic Forum's *Global Risks Report 2011*[5] put it: "the conditions that make improved global governance so crucial – divergent interests, conflicting incentives and differing norms and values – are also the ones that make its realisation so difficult, complex and messy."

Whichever way we look at the current landscape of structural drivers and macro trends, it is clear that we need to redefine what progress looks like. With major systemic changes at play, alongside seismic advances in technology, there are enormous challenges ahead but they also represent a wealth of fresh opportunities to reinvent the way we do business.

In 1962, Thomas Kuhn's influential book *Structure of Scientific Revolutions*[6] described how deep crises often lead to a paradigm shift, disturbing concepts of continuity and orderly progress by shaking the foundations of our worldview. It is my belief that such a paradigm shift has been happening since the beginning of this millennium because of global financial, geopolitical, and structural turbulence, along with major environmental challenges. What best defines our current status quo is the term "creative destruction," popularized by the Austrian-American economist Joseph Schumpeter back in the 1950s.[7] Schumpeter's change-focused economic theories explored how the dynamics of innovation and entrepreneurship in society can harness a more democratic way of creating economic value. The impact of this has resulted in a breaking down of traditional capitalist systems to give way to new economic models driven by collaboration, networks, and the social good. All too often, new thinking emerges as a result of a crisis rather than a carefully planned strategy. As we adjust to a new reality and establish fresh parameters to guide us, the single most important message for businesses is to refocus their message for the purpose of empowering people and communities alike.

## Small means agile and manageable

Possibly because of current social structures, there tends to be a belief that responsibility for the direction of society rests with "the others," but we all need to actively participate in shaping a more inclusive future. To do this, we must agree in principle how society's welfare systems should operate. Revisiting Schumacher's philosophy, being small might be a key survival strategy in the future. Small means agile and less at risk in a volatile and changeable world, but it also means manageable, because when people can relate and connect on a more localized level, they feel greater involvement and thereby responsibility. This feeds into a positive cycle of collaboration and caring – which increases personal happiness.

For nations and businesses alike, it is essential to nurture relationships, and this must be driven by an agenda of human-centric values, in which the state serves citizens and not the other way around. In international comparisons, not least the World Economic Forum's *Global Competitiveness Report*[8] (Figure 6.1), the Nordic countries – Sweden, Norway, Denmark, and Finland – invariably rank at or close to the top in areas such as equality and social inclusiveness; and yet one of the most cherished and often stated values of the Nordic way of life is individual autonomy – a sense of personal control rather than a feeling of being controlled. As the World Economic Forum's report *The Nordic Way*[9] says: "economic policies that cater both to our desire for individual autonomy and our need of community and security can be remarkably successful."

### *The Nordic Way*

There are three important lessons to be learned from the Nordic approach:

- Promoting autonomy can lead to greater social cohesion. People – especially women – feel empowered when gender-equal educational systems, individual taxation, universal day care, and anti-patriarchal family laws are embedded in policy.

- In the Nordic countries, social trust, confidence in the state, and relative equality coincide, and there is also a high degree of inclusion of citizens in the governance process.
- When grassroots groups are supported via effective and inclusive civil society networks, citizens' capacity for positive contribution, social responsibility, and self-realization are, in turn, enabled and developed.

Perhaps the secret of their success is that the Nordic countries have one of the best governance records. There is a long history of transparency, and developed systems of e-government mean that in Sweden, for example, everyone has access to all official records, and Norwegians can access the

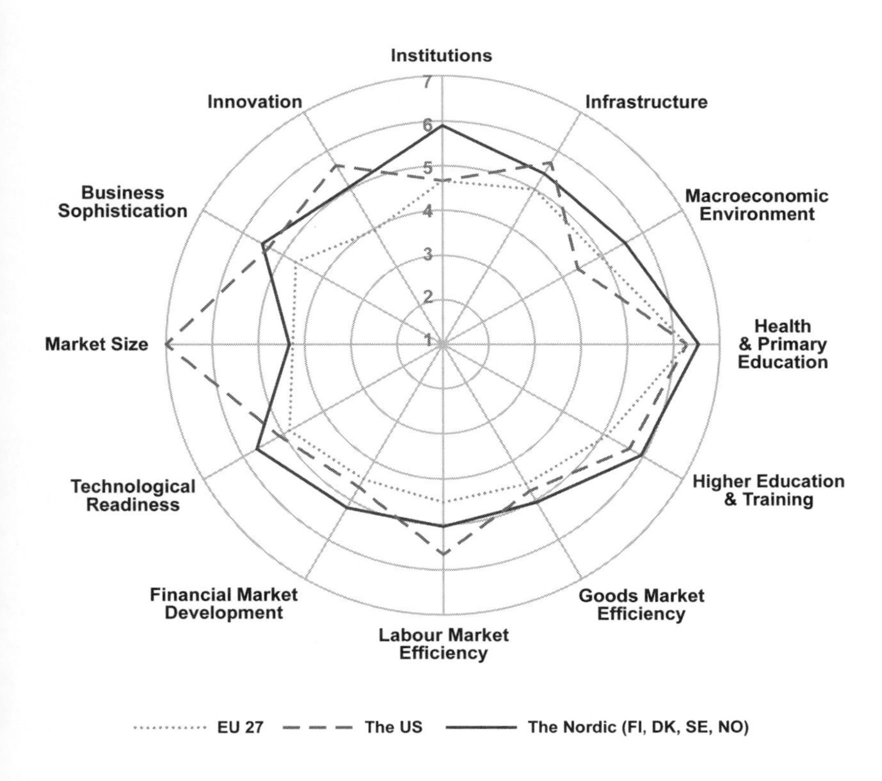


FIGURE 6.1 **Nordic competitiveness**: The Nordic countries rank at or close to the top in most areas
*Source*: Kjaer Global
*Data*: World Economic Forum Global Competitiveness Report 2010–2011

wage records of their fellow citizens. Danish government transparency has earned it the title of "least corrupt nation in the world" from Transparency International.[10] In *The Economist*'s report The Nordic countries: the next supermodel,[11] it was suggested that: "Smallish countries are often in the vanguard when it comes to reforming government." It concluded that: "The main lesson to learn from the Nordics is not ideological but practical." So it seems that being pragmatic and agile may be the best ways to remain focused on people and the greater good. These appear to be key to Nordic success – offering lessons that can usefully be adopted by business.

## The networked society

Our society is increasingly made up of networks of people, services, and environments that all have a purpose according to the relationships they enable. Governments and businesses, as well as people, are already a part of a greater information system, and this will be an added asset that has the ability to capture, communicate, compute, and collaborate around stored information and data. We will all have to define our position within this ecosystem of hyper-connectivity, and for business it's a key element to factor into any planning process, which poses this question: What role will the IoT play in the medium- to long-term future of my organization? As explored in Chapter 5, as many as 50 billion devices could be connected to the IoT by 2020.[12] This will have a profound impact on future society and our daily lives, and at this early point of the digital age, we haven't even started to grasp the full potential and economic range of the Internet.

The key task right now is to predict where the real value will lie, and who will be willing to pay for it. These "smart" assets have the ability to make processes more efficient and secure, give products and services new capabilities, and spark novel business models. To do this, we must reconsider the role of R&D, engineering, and design methods. If we start by putting the four Ps at the core of this process, we can begin to identify the real needs of all stakeholders to create more meaningful products and services. Tomorrow's designers and engineers need to embrace whole-brain thinking to decode complexity; their role will be to transform vast networks into open enabling systems of interacting, interrelated, or interdependent elements.

## Human touch interactions

Many voices, including scientists and technology gurus, are more than willing to tell us what technology can do in the future, but often they forget to factor people into the equation. With every exciting technological vision of tomorrow, it is worth reminding ourselves of a practical reality; however digitalized our world may become, human beings remain analogue. Our path to a true "value-based" integration of the two worlds will cause friction and disruption; nonetheless, our digital reality is moving along a rapid trajectory that will eventually deliver more organic, intuitive, and "human touch" interactions – all with a promise to empower and better our lives.

Currently, value on a societal level is largely assessed around collaborative tools, increased transparency, open source, real-time sharing, knowledge exchange, social networks, and living, breathing digital communities. The impact of this reality is already evident in the rise of dialogue-driven society and business models driven by disruptive innovation – factors that will become even more prominent over time. Companies and organizations that find a way to manage the data deluge and help us navigate complexity will win our loyalty. There is already a huge shift from physical ownership to the sharing of virtual services, driven by a generation used to open source and free downloads. They want to be involved in the process of developing products and services – they demand dialogue.

## Using design thinking for reflection

Design thinking is a useful way to manage some of the processes and leaps required to deliver a truly integrated dialogue-driven tomorrow that satisfies people's real needs. In design thinking, reflection is an essential part of the process of finding solutions; it requires us to step back and take a broader view of the problems or challenges at hand. Any problem is viewed as part of a larger interconnected system, one where the solution is likely to require a holistic understanding to connect all the individual components. This is why knowledge can be seen as a system of dots that must be connected in order to envision and make sense out of the whole.

The process tends to be open-ended with an intuitive structure, and therefore requires a cognitive thinking style that assesses the challenge in the most effective way. Basically, two main modes are required in design thinking: open thinking (intuitive and visionary) and closed thinking (rational and analytical). The design-led process is, therefore, an excellent example of multidimensional thinking, in which we switch from open to closed mode, depending on the challenge to be solved. This typically generates several possible solutions to one problem – perceiving information as "one big picture" – and the process as exploratory. Once the bigger picture has been explored, it's important to narrow down options. This requires analytical skills; and rests not just on understanding the problem, but on framing the criteria for evaluating possible solutions in a rational or closed mode, using systems thinking and several iterations to select the final path or choice.

SUMMARY: The people factor

- Purpose-driven leadership requires us to put people at the heart of the organization, refocusing brands so they empower people and communities.
- Being small could be a key survival strategy, as it means agile and manageable on a localized scale, giving us a more direct sense of our contribution to society.
- In the post-capitalist future, organizations will need to make a positive contribution, on all levels in society, through their activities.
- Companies have to consider human exchange not simply in the context of business, but in terms of a new economic paradigm of betterness.
- R&D must consider the impact of the IoT and plan around people, as open system networks that deliver organic, intuitive interactions with a human touch.
- Reflection is key to design thinking – tackling future problems – a process of connecting the dots to see the bigger picture before narrowing down to final decisions.

## How to build the business case

It is crucial to contextualize tomorrow's world to get a clearer idea of how change might manifest and influence your organization by studying the behavior of people, consumers, and competitors. Trend mapping helps you understand your current business climate and its future risks and opportunities. When we consider evidence for the value of trend management to business, IBM's 2010 global CEO study[13] is an interesting case in point. IBM interviewed 1,500 CEOs globally and found a general consensus on three key points: complexity is escalating; enterprises are not equipped to cope with this complexity; creativity is now the single most important leadership competency for all aspects of leadership, including strategic thinking and planning. The trend management system meets these three crucial needs for modern business because it is designed to navigate complexity and assist organizations in managing the overload of data and insights. More than that, it assists with the creative process by inspiring input, consideration, and feedback throughout the process. This method of checks and balances enables your organization to imagine and test out theories and ideas before translating them into business or innovation strategies.

Once you have established a need for trend management, it is essential to make a compelling business case with well-defined, desired outcomes to ensure buy-in from all levels within your organization. This is why it is so important to start with a clearly defined purpose – an end goal – and work carefully to that plan. The other way to get buy-in from the start is to focus on the by-products of the trend management process. By nurturing creative thinking, your organization develops an ecosystem of inspirational strategic initiatives that will strengthen and cultivate internal innovation culture, enabling people to play to their strengths – and when people thrive, the organization generally thrives too.

Indeed, what is so rewarding about working with trends is that once you start to apply your insights in a strategic framework, they will act as a GPS for navigating complexity – and you will realize that trend mapping is absolutely critical to all business decisions, because exploring socioeconomic and cultural drivers enables you to visualize how they might impact the business environment and the people within it. Further, it enables you to discover how designing and planning things differently might improve the ecosystem in which you operate, by introducing new parameters that balance sustainability and prosperity.

By combining several society drivers and macro trends from the Trend Atlas, you can create powerful scenarios. These inspirational future narratives are not predictions, but rather explorations of potential futures, and a logical starting point for planning ahead. The evidence presented throughout this book clearly shows that conventional methods of qualitative and quantitative research and traditional data gathering just don't cut it anymore. This is echoed by Mikael Ahlström, Swedish serial entrepreneur and a partner at Hyper Island (an educational company specializing in real-world industry training using digital technology), in a 2011 interview:[14]

> The term research should be renewed, since the best way to keep up has less to do with searching in old material and more about finding a way to tap into the constant flow of information, as well as engaging in conversations with your customers and industry leaders.

A coherent trend management strategy is one clear way to take on board Ahlström's assessment, in that it enables you to anticipate and capitalize on change by creating a bridge that enables effective communication across all departments within your company. It becomes the meeting ground between the technical and creative departments of your organization by providing a systemized way of using data, insights, and inspiration that are aligned with a specific consumer mindset or target market. Perhaps the biggest benefits of trend management are that the design, branding, and marketing of your products become much more efficient and profitable when you understand the drivers shaping the market in your planning phase.

Weaving in storytelling, in the form of viable scenarios, assists throughout the process by providing inspirational narratives that get under the skin of consumer behaviors to consider the product and lifestyle preferences that will influence future decision-making processes. According to Alex Osterwalder, co-author of the bestseller *Business Model Generation*,[14] all too often the focus is on balance sheets and business rivals at the expense of product. Converting from a business culture where the bottom line takes priority may sound like a tall order – especially in a tough economic climate – but, as he says:[14] "We focus too much on competition and financial objectives rather than being value and customer centric."

## Experimentation and disruptive innovation

Osterwalder adds that experimentation through disruptive business models is crucial to future business innovation, but it should not be at the expense of the current game plan:[14] "Keep focus on the core business while building teams that experiment with business tools for tomorrow. This is a challenge, but major companies must do this to maintain number one market position."

Creativity and experimentation are requirements to remain on top, which is why trend management's use of multiple methods is so valuable for considering the risk and opportunities ahead. Research that incorporates qualitative and quantitative approaches – alongside interviews, workshops, polls, online interaction, and feedback – fuels the process of developing fresh ideas and new business concepts that address the needs and wants of tomorrow's people.

When it comes to true innovation, it takes courage to step outside your comfort zone of tried-and-tested processes, but it is absolutely crucial. In an increasingly disruptive business environment, tomorrow's ideas might be radically different to today's – and that means they need to be envisioned. Osterwalder believes that all too often business leaders invest in innovation of more of the same, rather than inventing the business that customers want to buy into. He singles out Skype as an example of a truly innovative and customer-centric vision, noting that it was:[14]

> very similar to established telcos but with a very different business model, not needing a network and thus zero capital expenditure. The idea was a disruptive business model and started to compete with a new model and a new economy.

Leaders have to engage a creative mindset when preparing for the challenges of a disruptive business environment. In this context, organizations must hold a people-centric view of the user, not as a passive spectator and consumer but as an active creator, actor, and even producer. Brilliant innovations – such as Skype – revealed a need we didn't know could be fulfilled. Today, we take it for granted that we can speak to people around the world for free – revolutionary yesterday and taken for granted today. So the real question is: What comes next? This is where creating a Trend Atlas to support scenario planning comes into its own, in that it enables us to convert weak signals into innovative and disruptive business ideas by tuning into potential new wants – that people themselves haven't yet thought about.

### SUMMARY: Building the business case

- Trend management helps businesses navigate increasing complexity and apply creativity by designing and testing disruptive ideas before they are translated into strategies.
- To make the business case and ensure buy-in, it's important to focus on potential outcomes and this makes it important to have a clearly defined purpose at the start.
- A by-product supporting the business case is that the trend management process helps develop an internal innovation culture, enabling people to play to their strengths as well as acting as a bridge across departments.
- In an increasingly disruptive business environment, tomorrow's ideas might be radically different to today's – and that means they need to be envisioned through an iterative process involving scenario planning.

## Recognizing contradictions and transitions

Making sense of the complex and, at times, contradictory change drivers at play in our society and decision-making processes can be a challenge. Here, the Trend Atlas assists by enabling a clear overview and closer scrutiny of the potential influences that may lead to new behavior. It does this by focusing on and factoring in lifestyle patterns and influences that determine actual choices, as opposed to simply focusing on aspirations. This is important because, when it comes to people talking about their lives, their dreams, and their hopes, we have to remind ourselves that what they say and do may differ – a situation known as cognitive dissonance, as highlighted in Chapter 2.

We might understand the sociology of people, but we also need to include the sociology of things, places, and experiences to see how everything is interconnected. Therefore, sociological imagination[15] is essential because it gives us a much broader context in which we can detach ourselves from a situation and view the world holistically, allowing for contradictions and transitions in time to move our focus into a whole new dimension.

As explored earlier, future scenarios about society, businesses, and people deepen our understanding of issues that concern all of us, such as declining trust in institutions, diminishing resources, urbanization, aging populations, and interconnectivity. But they can also help decision makers reconcile uncertainties, contradictions or transitions, such as how political change in one part of the world might impact the rest of the world.

Shell has been using scenarios since the early 1970s as a part of its business strategy to navigate uncertainty and volatility, and to make better decisions. This can help future-proof an organization in a volatile market or, as the company puts it:[16]

> Organisations using scenarios find it easier to recognise impending disruptions in their own operating environment, such as political changes, demographic shifts or recessions. They also increase their resilience to sudden changes caused by unexpected crises like natural disasters or armed conflicts.

Over the years, governments, academics, and business leaders have looked to the Shell scenarios for inspiration to help deepen their understanding of how the world might appear decades ahead. This has filtered down so that more and more businesses now make it part of their strategic practice to make sense out of the medium- to long-term future and navigate change with confidence.

## What is a scenario?

As already stated, the primary purpose of a scenario is to contextualize the future by considering socioeconomic and cultural trends, while also considering the human scale and attempting to detect otherwise hard-to-spot behaviors and patterns. Ideally, you start with a vision of how people, as citizens and consumers, are likely to be influenced by political, economic, social, and cultural factors, as well as considering the impact of technology on our cities, transportation, buildings, and infrastructure. What develops from this is a scenario – a visual collage or a written synopsis of a series of events, influences, and trends. It acts as a storytelling framework for exploring the bigger picture of selected macro trends and then examining more nuanced details with supporting micro trends. The resulting composite enables you to translate ideas and vision into a tangible action plan for your organization and create a common road map of the future you want to see.

Scenarios are developed to map out how data and insights might evolve in a particular context and can also be expanded into fully developed concepts or strategies. As a possible profile of the future, they can be helpful in pitching ideas and concepts, or in testing and comparing a variety of possible outcomes to improve creative thinking and develop projections for the most likely or lucrative outcome. When working with several scenarios at once, all should have the same time frame and/or similar likelihood in order to decrease bias and make them truly comparable.

Developing scenarios in a future storytelling format means they are narrated in an imaginative and memorable way; however, this approach has a logical framework underpinned with data and evidence that gives the audience a better sense of how changes could impact on people's lives in the future. There is no one perfect way of writing scenarios, but the most robust starting point is to build them around personas – what we call "core typologies" – as this will make them feel more real. Underpinning all our scenarios is the 4P philosophy, in which we develop strategies that consider people, planet, and purpose alongside performance.

## Consumer mindset mapping

The Mindset Map (Figure 6.2) is a tool developed to deliver meaningful insights into how the influences from the Trend Atlas impact people's lifestyle choices and value universe. By linking key trends and associated values, we can narrate the core typologies in order to facilitate a deeper understanding of people and their lifestyle preferences, while also highlighting the challenges and opportunities that need to be addressed. The framework for these personas evolved over the years, as a result of working with a diverse range of industries and endeavoring to meet clients' need for more information about their end users and tomorrow's people, either as citizens, consumers, or employees. Core typologies and future scenarios must be understood according to prevailing micro and macro trends, as well as geography, local culture, and economy. The method has proved to be an effective way of imagining tomorrow's people and establishing a balanced worldview in a digestible and easy-to-communicate snapshot format. The information may be presented in brief note form or developed into detailed profiles to understand and envisage likely viewpoints and life choices.

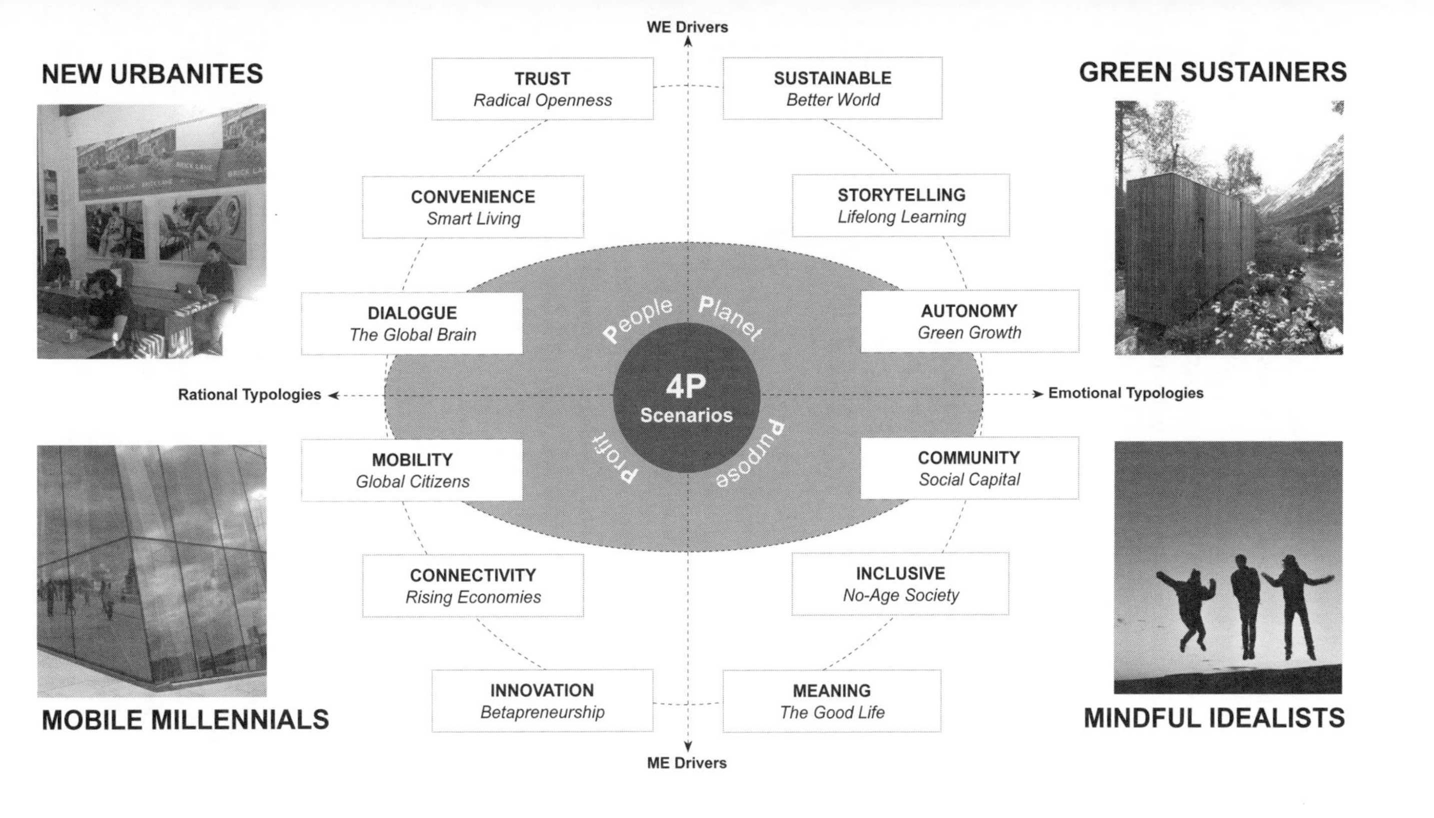


FIGURE 6.2 **Mindset Map and the 2030+ scenarios**: Snapshot of tomorrow's people showing four core typologies
*Source*: Kjaer Global

### The core typologies and scenarios

Figure 6.2 above illustrates a snapshot of tomorrow's people, showing core typologies to visualize how the trends they relate to most closely inform their preferences and values. Figure 6.2 differentiates between rational and emotional types on the horizontal scale and ME and WE types on the vertical. Rational and emotional types are shorthand terms for people led by the heart or the head (pragmatists and idealists). ME people focus on life according to themselves, while WE people are more engaged with the group and community. While such terminology is a shorthand, it's worth noting that dividing key characteristics into typologies gives a lean overview of people and future scenarios – offering an at-a-glance mindset comparison. Naturally, people are not fixed "types" in real life, instead they tend to be dynamic and overlap depending on their situation or needs. People form groups across conventional borders, and therefore must be understood in terms of their affiliations, culture, and value universe. The Mindset Map enables us to view the world through their eyes in order to create relevant and engaging narratives. In the following future scenarios, we have created a handful of inspirational narratives built on the back of the interlinked structural drivers and macro trends explored in previous chapters.

The four scenarios we explore are: Mobile Millennials, New Urbanites, Green Sustainers, and Mindful Idealists. As evident in the following narratives – inspired by an idea we originally developed for the book *Vision 2030: So leben, arbeiten und kommunizieren wir im Jahr 2030*[17] – there is no doubt that we are entering a world where new values and norms will emerge.

Underpinning the development of the four scenarios is a robust evidence base from the trends in Chapter 5, supported by desktop research, expert interviews, workshops, academic papers, and literature reviews, organized around five research challenges. Each scenario has a Trend Index to support the profile with keywords and core trends highlighted, plus a visual storyboard to support the three core headings and form a 360-degree view. Core drivers summing up the scenario are found in the How to Spot Section, under the headings: value universe, work and third space, home and leisure, and products and services.

## Scenario 1: Mobile Millennials

Mobile Millennials (Figure 6.3) are a generation with a rational and ME-centered worldview. They have matured into a wider group of creative classes, influencing society and culture on many levels. These Global Citizens thrive in urban settings defined by their high levels of innovation and

FIGURE 6.3 **Mobile Millennials Trend Index**: The key trends impacting the scenario on a 1–5 scale
*Source*: Kjaer Global

talent density. The Rising Economies offer a fertile ground to develop their passion for Betapreneurship and they help to create new power hubs as they continue to grow and flourish as individuals.

## A world without boundaries

Worldwide, over half the population have joined the consuming classes, impacting geopolitics as the axes of power have shifted. China and India are the world's largest Internet hubs, presenting myriad fresh work and life opportunities for the hyper-mobile Global Citizens. Agile development and an entrepreneurial mindset build flourishing economies, and are supported by government through business start-up schemes and entrepreneurship programs within schools. Many large corporations are more influential than governments in driving positive change because they nurture local communities and the platforms needed to deliver on the 4Ps. Smart organizations know that intrapreneurship programs – nurturing entrepreneurial thinking from within – are crucial to attracting a diverse and talented workforce with the vision to find innovative solutions to tomorrow's business and society challenges.

## Cultivating a "smart culture"

The Mobile Millennials have grown up and are reshaping the worldwide talent pool. Most are tech-savvy and informed individuals, self-styled media experts, curators, librarians, publishers, and journalists – and all connected via the Global Brain. This is paving the way for an open-minded, collaborative audience seeking intelligent communication and diverse experiences in real time. Already, it has inspired new working styles and consumption patterns, where companies give "carbon reduction" incentives to remote workers. These scorecards can be exchanged for local gym and dinner club cards to support healthy living away from work, as this is a vital part of new CSR guidelines across the world. Increasingly, new "sharing economies" – especially in Asia – fuel flexible and sustainable models, as people choose access to goods over ownership.

## Connected living experiences

Technology has finally reached a true convergence stage, creating a wide spectrum of human touchpoints. Mobile Millennials gravitate towards the most liveable cities that display high standards of smart connected living, education, and civic values. This empowered generation want meaningful experiences – urban connectivity, modern architecture, hybrid cuisine, and culture-quest travel – seeing this as their opportunity to create one-off experiences. As multitasking digital natives, they forge friendships, work and business relationships across the world – to them, social storytelling is not only multichannel, but also intuitive and responsive, adapting according to media and environment. This dynamic generation seek more than entertainment – experiences must be heartfelt and reflect personal needs and values – so networks and brand interactions are an anchor bringing new modes of belonging (Figure 6.4).

## HOW TO SPOT MOBILE MILLENNIALS

### Value universe

- Seek flexible and meaningful options promoting sustainability
- Know that cultural and Social Capital support personal development
- Believe they can be the change they want to see

### Work and third space

- Not confined by space – form alliances across borders
- Challenge current notions of work–life balance
- Thrive in the intersection between disciplines

### Home and leisure

- Create smart connected habitats to foster belonging anywhere
- Participate in self-monitoring as part of a healthy lifestyle
- Embrace digital-centric platforms and immersive cultural experiences

### Products and services

- Expect convenience – real time and "on demand" as standard
- Storytelling must resonate with their immediate environment and culture
- Look for brands supporting their worldview and notions of belonging

# MOBILE MILLENNIALS: Rising Economies/Global Citizens/Betapreneurship

A world without boundaries

Connectivity

Informed

Cultivating a "smart culture"

Disruptive innovation

FIGURE 6.4 **Mobile Millennials visual snapshot**: This storyboard is a visual summary of Mobile Millennials' key trends and values
*Source*: Kjaer Global

## Scenario 2: New Urbanites

These rational WE-centered New Urbanites (Figure 6.5) are family-oriented social connecters who enjoy urban living in smart cities – enabled by Radical Openness and the Global Brain. They actively engage in civic pursuits and are the backbone of the sharing economy for the greater good of all. Champions of Smart Living – whether it's converting urban wasteland into allotments or starting local walking/biking clubs – they are tireless ambassadors of community activism and values that make cities thrive.

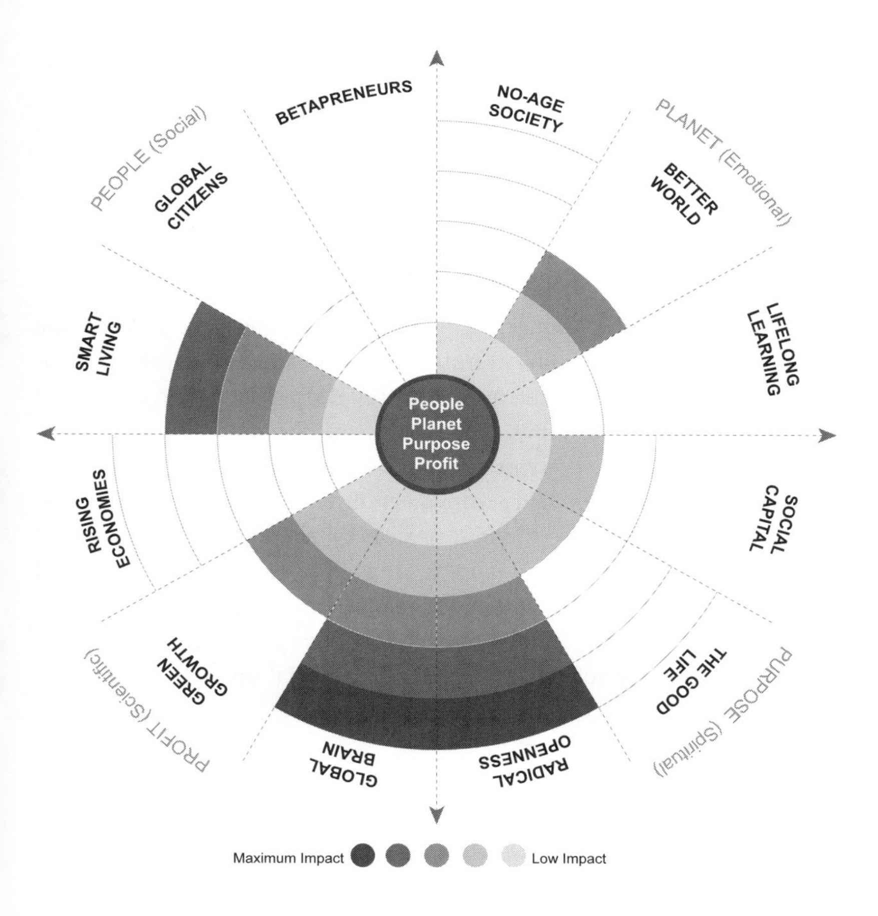


FIGURE 6.5 **New Urbanites Trend Index**: Indicates the key trends impacting the scenario on a 1–5 scale
*Source*: Kjaer Global

## The "smart grid" society

Urbanization continues to be an overriding global theme, and utilizing surplus space for growing food, harvesting gray water, and sourcing energy is an essential factor in creating sustainable environments. Thinking cities are based on a smart infrastructure that enables civic-minded urban dwellers to thrive in population-dense settings. Now instrumental in saving time and resources, an intelligent grid connects the city itself with government, businesses, people, homes, transport, and devices to create a truly collaborative society. Transportation mainly consists of intelligent driverless electric units, navigating via a smart network to reduce congestion and pollution. Urban centers are largely walkable, with fast-track bicycle lanes, and commuting from suburbs is enabled by car-sharing schemes and cleantech public transport. Together, these factors contribute to more liveable cities.

## Collaboration-driven communities

New Urbanites believe that collaboration lies at the heart of thriving societies – sustaining families, uniting communities, and making life meaningful. This philosophy translates into greater wellbeing, sustainability, and savings – even the most skeptical people have been persuaded to join the sharing economy as it makes sound economic sense. These smart connected citizens are quintessentially creative and frugal consumers, demanding value, quality, convenience, and authentic experiences, all in one package. Home sharing, flexible car-sharing schemes, canteen cooperatives, bartering, and swapping are their way to reconnect and find harmony and belonging in virtual and real-world communities. Consumption is optimized and they do their bit as individuals by opting for "less but better" – buying collectively through local social shopping schemes or opting for tailored, reduced versions of products.

## Emotionally responsive technology

These collaborative individuals choose sustainable new-build or energy-efficient refurbished properties, but always connected to the Global Brain. In their smart home, emotionally responsive alert systems are standard, Internet-connected bathrooms are fitted with health-diagnostics equipment and real-time kitchens connect fridge and cooker with analytics software to optimize behavior. These systems deliver personalized self-monitoring support to the whole family, promoting individualized, balanced lifestyles. New Urbanites recognize that being part of the connected world is essential to managing daily life but can blur the boundaries between public and private space, challenging the work–life balance. They believe in Radical Openness and see the trade-off in less privacy as worth it because it delivers redefined wellness and safer, more self-sustaining communities, bringing huge benefits to everyone (Figure 6.6).

## HOW TO SPOT NEW URBANITES

### Value universe

- Participate in civic activities to build local Social Capital
- Drive the sharing economy and subscribe to "less but better"
- Believe that Radical Openness creates stronger communities

### Work and third space

- Form new businesses based on bartering and sharing ideas
- Find ways to utilize urban spaces and optimize resources
- Support community networks and real-time democracy

### Home and leisure

- Pioneer new green initiatives in the urban environment
- Embrace connected Smart Living solutions using new technology
- Seek redefined wellbeing to stay active for longer

### Products and services

- Demand value and values – unsustainable is not an option
- Use their networks to collaborate and share with others
- Aspire to experiences that optimize lives and wellbeing

FIGURE 6.6 **New Urbanites visual snapshot**: This storyboard is a visual summary of New Urbanites' key trends and values
*Source*: Kjaer Global

## Scenario 3: Green Sustainers

The emotional and WE-centered Green Sustainers (Figure 6.7) are at the vanguard of considered consumption, viewing Green Growth as the only way to optimize natural resources to leave a positive legacy for future generations. They cultivate a Better World by harnessing new technologies to lead positive change, but are also passionate supporters of artisan initiatives to cultivate Lifelong Learning that supports local community growth. Empathic and innovative, they influence others through their resourceful and mindful approach.

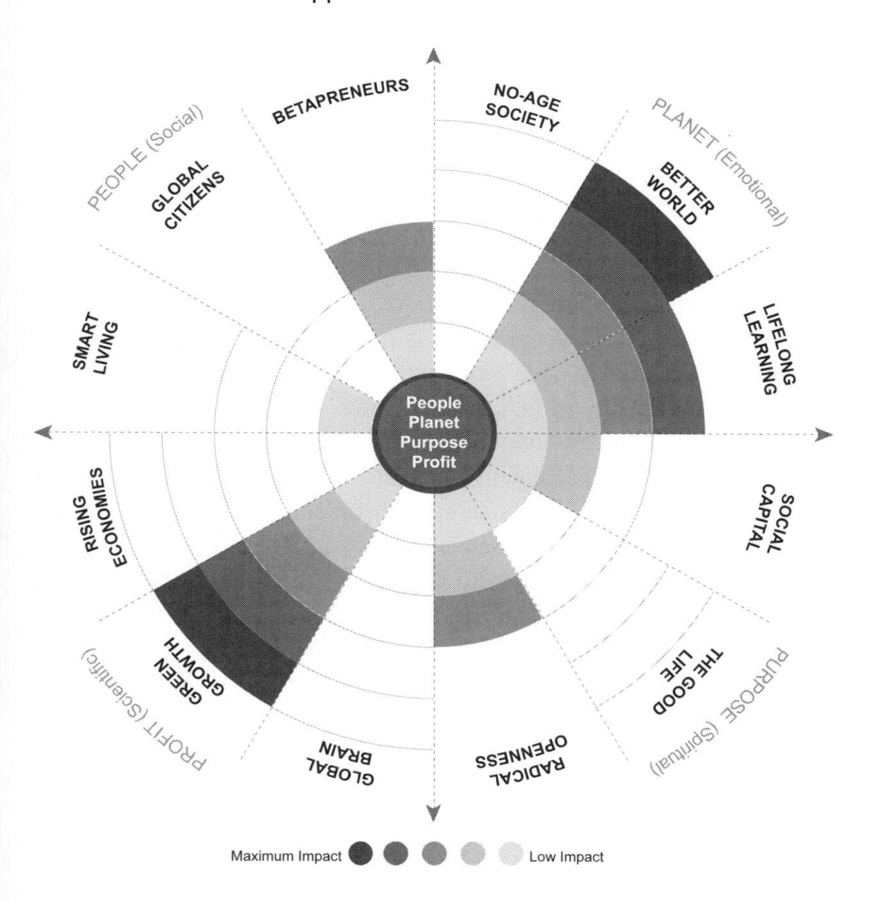


FIGURE 6.7 **Green Sustainers Trend Index**: Indicates the key trends impacting the scenario on a 1–5 scale
*Source*: Kjaer Global

## The Circular Economy

With steep growth in population and continued concern about climate change and resource shortages, circular and cleantech economic principles have been adopted across most nations – with Scandinavia and parts of Asia exporting their know-how to assist other regions. Female presence in the boardroom and in government is key to forging valuable new partnerships and alliances across traditional industries and borders. Businesses are driving systems thinking as a solution to global challenges and have recognized the value of 360-degree practices to grow a meaningful brand index. Openness on sustainability issues is key for organizations that want to demonstrate they care – and Green Sustainers have been instrumental in rewriting the ground rules for business success to focus on preservation of natural resources.

## Everything is connected

The merging of products, services, and experiences has meant reconsidering the way we consume, with information about everything from origin and ingredients to carbon footprint and production methods being provided as standard. This openness is partly as a result of information campaigns demanding accountable and forward-thinking practices on social responsibility. Green Sustainers are among the new breed of resource-strong social entrepreneurs playing an active part in shaping small but significant maker movements – a cultural phenomenon that emphasizes the importance of storytelling. Rediscovering the story of people, process, and planet is central in reviving "crafted with care"; human-centric processes demonstrating empathy for local culture and context play an increasingly important role in brand innovation and success.

## Considered consumption

Profoundly optimistic in their betterness agenda, Green Sustainers choose, retrofit or build smart, low-carbon homes, maximizing efficiency by combining modern technology with traditional, built-to-last approaches. They engage in Lifelong Learning – be it growing their own produce, harvesting their own energy and selling the surpluses to the national energy grid, or being "fixperts" supporting a global repair community. With their love of provenance in food, furniture, fashion, travel, and art, they seek out "the real thing" and excel at sharing best practice, using technology and affinity networks to stay ahead of developments. For them, considered consumption means positive action to nurture and replenish the global village, so if they travel, it may well be on a volunteering holiday to help other communities grow their sustainability activities (Figure 6.8).

## HOW TO SPOT GREEN SUSTAINERS

### Value universe

- Partners with nature – sustainability is integral to their lives
- Values-based thinking: people, planet, purpose, and then profit
- Strive to balance natural, human, social, and economic capital

### Work and third space

- Choose empathic organizations with strong values and ethics
- Encourage diversity and an inclusive workplace
- Promote local growth through social entrepreneurship initiatives

### Home and leisure

- Replenish and rethink – reduce, reuse, recycle approach
- Support local green initiatives and cleantech solutions
- Prefer new artisanship, heritage, and cultural narratives

### Products and services

- Work towards a "circular" lifestyle and love fixing things
- Look for sound ethical foundation and transparency
- Maximize social networks for "good" using technology

**GREEN SUSTAINERS:** **Green Growth/Better World/Lifelong Learning**

The Circular Economy

Storytelling

Sustainable

modern farmer

feast!

EAT. DRINK. FARM

mexico, michigan, slovenia

Participation

Autonomy

Everything is connected

Considered consumption

FIGURE 6.8 **Green Sustainers visual snapshot**: This storyboard is a visual summary of Green Sustainers' key trends and values
*Source*: Kjaer Global
*Image*: Autonomy: "Mapping Design for Circularity," Credit: The Great Recovery Team at the RSA

## Scenario 4: Mindful Idealists

The emotional, ME-centered Mindful Idealists (Figure 6.9) look to the Good Life as the ultimate goal in everything they do. As drivers of a No-age Society, they champion a 5G workforce with flexible work schemes as a route to maintaining a healthy welfare system and Social Capital. They are catalysts of change – their idealism convinces others – and work to be in control of their lives and in tune with their community.

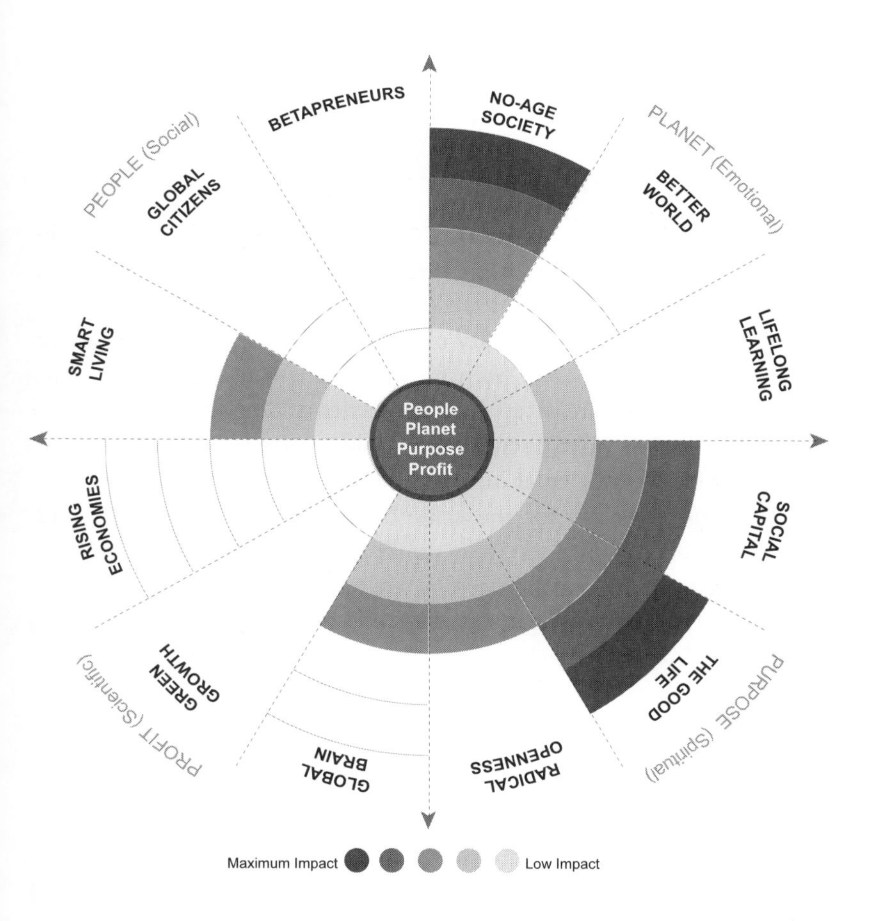


FIGURE 6.9 **Mindful Idealists Trend Index**: Indicates the key trends impacting the scenario on a 1–5 scale
*Source*: Kjaer Global

## Positive life experiences

Factoring in the total wellbeing of people, planet, and society, alongside traditional GDP, has become the economic norm across all OECD countries, with most nations getting closer to good life ideals – where GDW (gross domestic wellbeing) is defined by liveability, social mobility, and access to information. This supports the Mindful Idealists' aspiration, which is to live in a world where the goals of business and society are aligned. They are passionate supporters of inclusive approaches and expect total transparency from government and organizations, viewing this as fundamental to trust and strong community. This informed group support grassroots politics and favor alliances – it's important for them to express fresh ideas that make a positive difference at a local level and improve lives.

## The radical open society

The social and emotional benefits of any experience have made people-centered approaches the natural starting point for any new ventures. This goes hand in hand with increased collaboration and more integrated public services to improve quality of life. Mindful Idealists favor Radical Openness across the board, both for equality and to ensure symmetry of knowledge. They are scrupulous about demanding and checking the data organizations hold about them – always considering the ethical standpoint before they enter into any relationship. Once convinced, they are powerful advocates for the value of big data derivatives – including e-governance, education, and public services – as the convenience of digital citizen ID and interconnected quick response codes systems gives them more control over their lives and the opportunity to focus on personal goals.

## Inclusive holistic solutions

A more caring and flexible approach is helping to instil greater trust across society. As supporters of the 5G workforce, with their own tight-knit "No-age" network of friends and family, Mindful Idealists are early adopters of assistive technologies, viewing them as the route to remaining independent for longer. Equally at home anywhere in the world – and often choosing to work remotely – they seek out No-age organizations that offer knowledge sharing and meaningful experiences. Control is key, so they happily adapt to the latest robotics and technology, personalized to support their employment and lifestyle choices and free up time to focus on the things that matter. Their lives must have a purpose, so they are always active – working, participating in Lifelong Learning and supporting local arts and social initiatives (Figure 6.10).

## HOW TO SPOT MINDFUL IDEALISTS

### Value universe

- Aspire to the Good Life as the ultimate goal
- Align their values, actions and personal goals to live holistically
- Believe that openness and consistency build trust

### Work and third space

- Champion a No-age workforce spanning five generations
- Catalysts of positive change – focus on purposeful lives
- Independent minded and strong supporters of inclusiveness

### Home and leisure

- Motivated by wellbeing monitoring for balanced lifestyle choices
- Feel liberated by technology to focus on more important things
- Attracted to improved personal control and smart interfaces

### Products and services

- Look for a balance between tactile and intuitive interactions
- Want responsive open stakeholder dialogue
- Demand a clearly defined betterness agenda

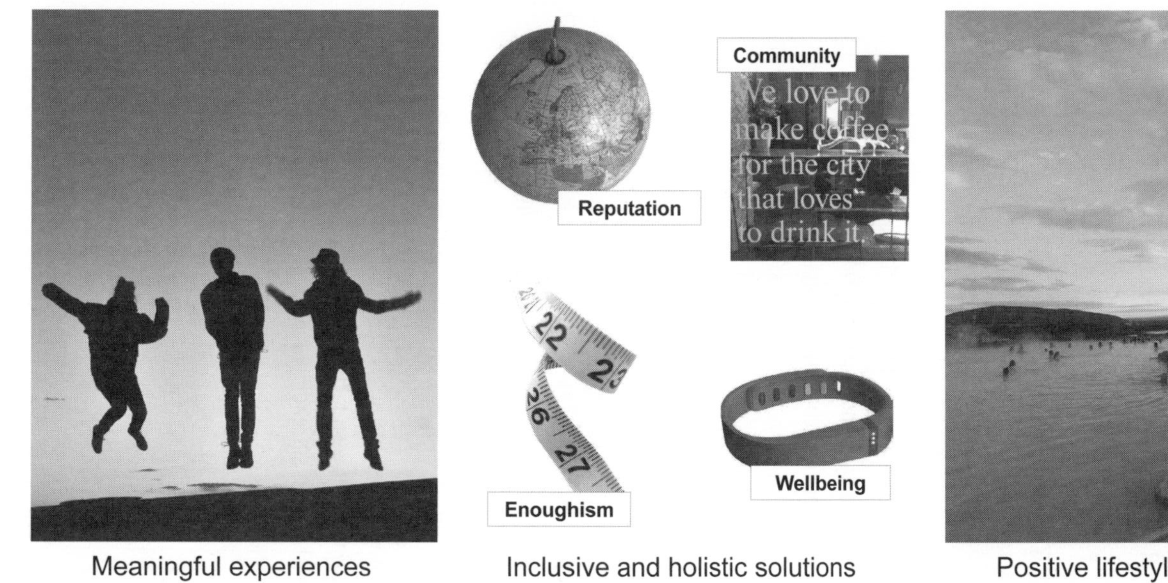


FIGURE 6.10 **Mindful Idealists visual snapshot**: This storyboard is a visual summary of Mindful Idealists' key trends and values
*Source*: Kjaer Global

## Trends across continents

Hyper-mobility has resulted in a patchwork society, where value sets and common interests unite people across geographical borders. But to make any trend-mapping exercise real, it is essential to allow for geographical differences in order to tune into the scale and impact of the trends in question – in a global and local context. Initially, you start by mapping out the most important global macro trends by looking from the outside in. Then you move these findings into a local context by looking from the inside out. Remember, when we speak about global influences, we must always consider how global scenarios play out at a local level in order to make them relevant in a living context, and ultimately take advantage of them.

### The importance of locality

To interpret the impact of a global trend in a localized setting, you have to look for evidence in supporting micro trends, and this can vary tremendously depending on the economic climate, culture, and environment impacting consumer behavior. Rediscovering the value and importance of the local village, as opposed to the global village, is the foundation of an inclusive and purpose-driven business approach. By understanding and engaging with communities at a closer level, you empower and influence people's lives in a positive way and become part of their story – and they, in turn, help to shape your brand narrative. Organizations that are already nurturing local cultural capital successfully do this by harnessing the strength and uniqueness of locality, rewarding businesses, communities, and individuals along the way. It's about thinking local in order to create a lasting cultural legacy and have relevance in the lives of the people you interact with.

This is not idealism, but a sound business proposition – one that organizations are investigating because it is a crucial driver to create Social Capital. Organizations and governments alike need to cultivate glocalization to improve people's lives, as this will deliver positive and enriching experiences, products, and services across society. There is a growing need for brands to embrace more human and conscience-driven values, but citizens will become increasingly proactive, in that they pick and choose, scrutinizing the company or brand as closely as the product. This is a concept I touched on in an interview in 2010:[2]

> Complexity and the excess of meaningless choice is still an unsolved issue in most developed western economies. Concepts simply must become people-centric, demonstrating empathy for the cultures they serve and respect for the context in which they exist. A challenge within this is achieving consistency throughout both intangible and tangible elements.

## Changing our perceptions

From politics and work to community and family, happiness and meaningful experiences are strong drivers. We seek emotional engagement to assist us in achieving personal fulfilment and quality in all areas of life. Developing a brand platform with the potential to act as a wellbeing facilitator in order to match the real needs of people will be of huge importance to future success. Obviously, rather than just one, there are many potential lifestyle scenarios for 2030, and the narratives presented here are meant to serve as an inspirational platform to instil an open vision of the future. These illustrative profiles, developed out of the Trend Atlas 2030+ (Figure 4.2), are designed to create engaging and believable narratives. Good scenarios have the potential to open minds, by changing our perceptions of the future in a profound way. We get rid of old assumptions by observing the world from a 360-degree perspective.

## Linking trend and behaviors

The Trend and Lifestyle Navigator (Figure 6.11) is a unique tool developed by Kjaer Global to synthesize macro and micro trends with behaviors and motivators by linking them to future challenges and opportunities. It provides a meaningful overview of how macro trends, lifestyle patterns, and value sets are intimately related to consumer behavior and needs, as it is essential to understand how trends are linked to people's lifestyle preferences when creating an innovation or business strategy. The Trend and Lifestyle Navigator provides a framework for summarizing the key insights from the scenarios by highlighting potential business opportunities. The tool is also designed to facilitate a synergy between divisions – design, branding, marketing, and retail among others – to ensure that future strategies are collectively understood and can be implemented successfully across organizational platforms and departments. The Trend and Lifestyle Navigator can be read horizontally and vertically. Horizontally, it illustrates how trends impact people's behavior, needs, and value universe. Vertically, it invites a lean overview and cross-comparison of consumer mindsets and insights.

Its intrinsic value is that it assists us in the main task for all impactful trend management exercises – unlocking the key to understanding tomorrow's world and bringing future opportunities to life by tapping into the emotional and rational landscape of the people you want to reach.

## Stories from the heart

In the challenging process of understanding how to best harvest, filter, and process information, let alone extract meaningful messages, companies can no longer rely on traditional models for market knowledge – the same old stories simply won't work any more. Businesses need to start thinking of themselves as facilitators, making it possible for people to build their own meanings into products and services. But how do we define a meaningful experience in a 21st-century context? While globalization and mass consumption appear to make the world increasingly homogeneous, at the same time we witness how people's search for meaning has intensified. Digital interconnectivity and social networks are the opportunity for people to create their own unique identity and personal brand image – it is them, not the companies they deal with, that take center stage and people are moving the debate about how we want to live and consume into new territory.

Purpose driven is a parameter impacting our lives around the clock, and we now expect it from our work, our social encounters, and our personal lives, as well as from the brands we interact with. The past few decades have seen a consumption revolution, with a growing demand for products and services that people can relate to and feel involved with – even fall in love with. The impact of this has been that people want the real thing in all areas of life, and happily mix the objects and experiences they collect into their own personal narrative. Meaningful experiences that bring inspired involvement and informed engagement must be embedded in the DNA of all thriving 21st-century organizations.

It is essential to "think big" to feel inspired, and this is best done by embracing change as an opportunity, with a mindset based on hope, rather than fear. By imagining what society will look like by 2030+, we can learn together and develop a common understanding of how we see

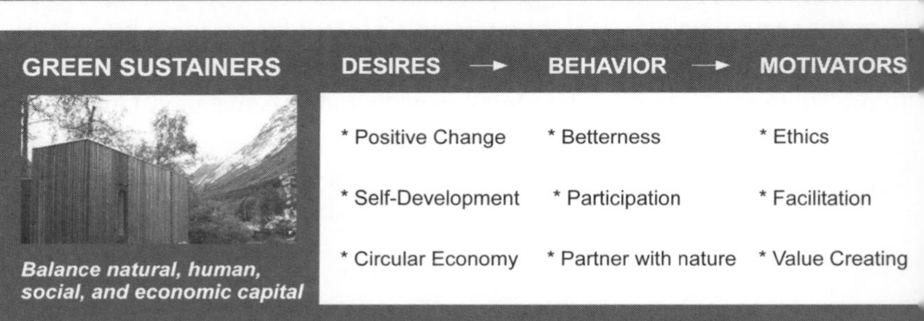


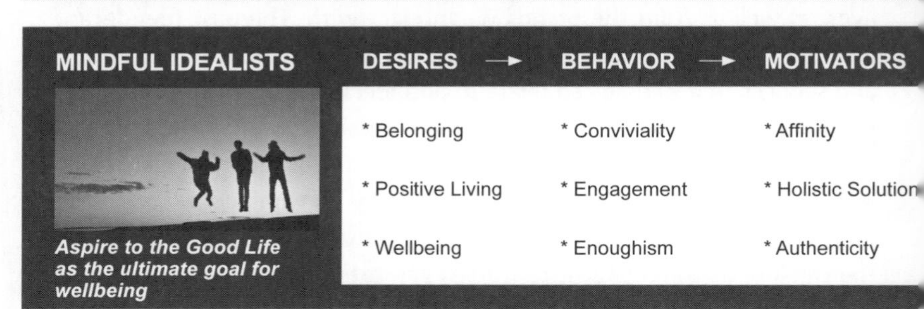


FIGURE 6.11 **Trend and Lifestyle Navigator**: This overview synthesize macro trends and behaviors by linking them to future opportunities
*Source*: Kjaer Global
*Image*: Green Growth: "Mapping Design for Circularity," Credit: The Great Recovery Team at the RSA

| MACRO TRENDS → | VALUES → | OPPORTUNITIES |
| --- | --- | --- |
| GLOBAL CITIZENS | MOBILITY | * *Growing, nurturing and rewarding new talent* |
| RISING ECONOMIES | CONNECTIVITY | * *Holistic approach with strong local presence* |
| BETAPRENEURSHIP | INNOVATION | * *Local social networks and grassroots approach* |

| MACRO TRENDS → | VALUES → | OPPORTUNITIES |
| --- | --- | --- |
| RADICAL OPENNESS | TRUST | * *Trust through consistency and accountability* |
| SMART LIVING | CONVENIENCE | * *Simple solutions through intelligent reduction* |
| THE GLOBAL BRAIN | DIALOGUE | * *Deep consumer knowledge to deliver 4P value propositions* |

| MACRO TRENDS → | VALUES → | OPPORTUNITIES |
| --- | --- | --- |
| BETTER WORLD | SUSTAINABLE | * *Clear goals and vision for leaders inspire stakeholders at all levels* |
| LIFELONG LEARNING | STORYTELLING | * *Invite people to share their own stories and experiences* |
| GREEN GROWTH | AUTONOMY | * *Adopting a holistic long-view for the benefit of people, planet and profit* |

| MACRO TRENDS → | VALUES → | OPPORTUNITIES |
| --- | --- | --- |
| SOCIAL CAPITAL | COMMUNITY | * *Happy communities = thriving businesses and people* |
| NO-AGE SOCIETY | INCLUSIVE | * *Education and mentorship on all levels* |
| THE GOOD LIFE | MEANING | * *Caring solutions that make a real difference* |

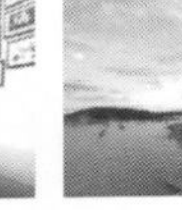

society evolving and how to best prepare for this. This is why the four scenarios in this chapter are a benchmark of the approach needed to imagine the future we want to create. Like all forecasting exercises, it carries certain challenges and risks but also reveals a wealth of opportunities. If you can step outside the comfort zone of the familiar world you inhabit to imagine the future, you are much more likely to make it happen.

---

## SUMMARY: Trends across continents

- Trend mapping begins at a macro level – from the outside in – and then adding the micro view by looking from the inside out. Always consider the local level of a macro trend to capitalize on them.
- Glocalization is crucial and is the way to generate Social Capital and have a meaningful impact on the lives of people and communities.
- Meaningful experiences and inspired involvement will ensure informed engagement – the brand fabric of any thriving 21st-century organization.
- The Trend and Lifestyle Navigator is a tool to synthesize macro trends and behaviors by linking them to future challenges and opportunities. It assists us in visualizing and understanding tomorrow's people.
- Businesses need to start thinking of their role as facilitators inviting people to build their own meanings into products and services.
- Making the future happen requires a common road map – because when we can imagine it, we are more likely to make it happen.

---

# CHAPTER 7

# Practical trend mapping: organizations

*A springboard for innovation, trend management also facilitates new ways of thinking about your core ambitions and business opportunities – and the best way to achieve them.*

Have you ever found yourself uncertain about how to move ahead because the challenges facing your organization or team made decision making almost impossible? For most people in a leadership role – be it in management and strategy or research, design, and development – that feeling of being overwhelmed by the complexity of the situation has happened at some point in time. The key question becomes: How do I see through the chaos and the contradictions to find clarity? At the same time, the fear of making an ill-considered decision is a very real one. Going in the wrong direction, because you did not fully consider all the evidence and viewpoints available, could compromise your position, your team, and even the future of your organization. In most instances, we do eventually get out of the "analysis paralysis" and start to formulate a plan, but there are tools to help overcome or even avoid these blocks effectively. The starting point is to stop worrying about the elements you can't control – the uncertainty – and instead activate the contemplation process by mapping out what you do know and can control.

During such periods, it is vital to think of the challenges ahead as a process, rather than a set of insurmountable obstacles. Crucially, in order to move ahead, accept that testing ideas that might not work initially is not a failure, but an intrinsic part of strategizing for success. The inventor

Thomas Edison is reported to have said:[1] "Genius is one percent inspiration and ninety-nine percent perspiration." He had thousands of unsuccessful attempts before developing the commercial light bulb prototype that changed his fortunes and history. A news reporter once asked him: "How did it feel to fail 1,000 times?" Edison reportedly replied:[2] "I didn't fail 1,000 times. The light bulb was an invention with 1,000 steps." This attitude is crucial to embed a mindset of creative leadership and betapreneurial thinking within your organization – one where a trial-and-error process is the guiding principle for navigating through the maze of complexity onto the firm ground of ideation and strategy formulation.

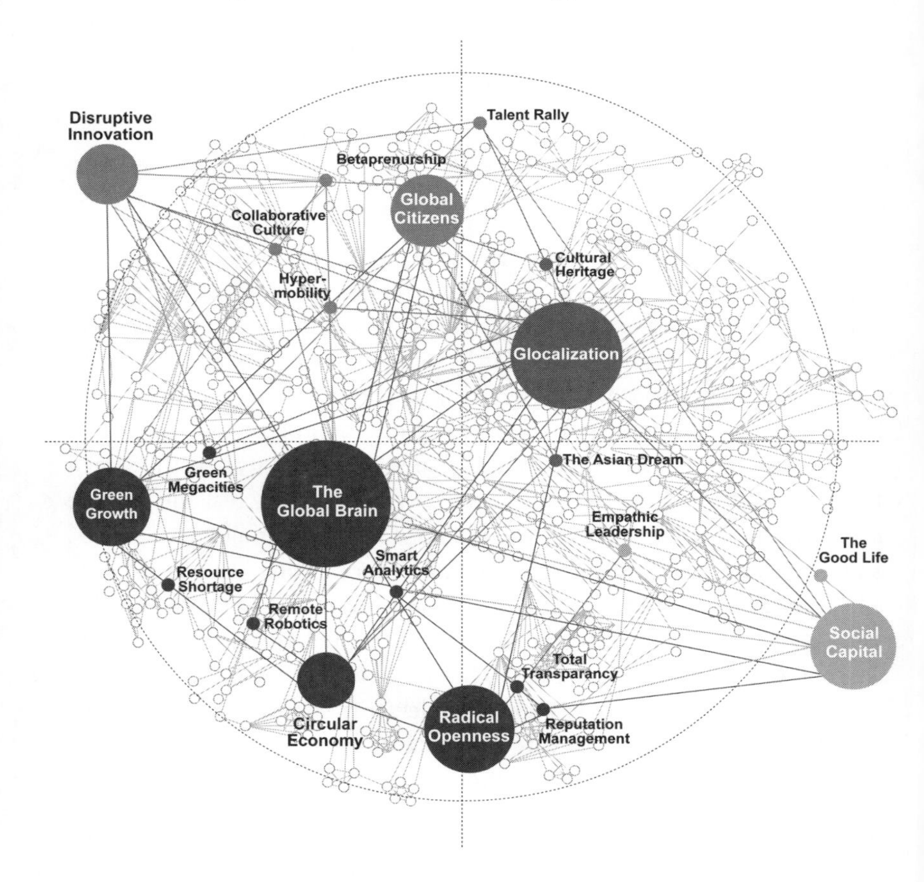


FIGURE 7.1 **Connecting the dots**: Clustering macro and micro influences gives us a broader and more meaningful perspective
*Source*: Kjaer Global

## From confusion to trend management

Viewing our reality as interconnected layers that need to be reorganized and reconnected is one way of coming up with solutions to a problem. When we isolate and observe trends and developments connected with our specific challenge, we start recognizing the opportunities that lie ahead – and this is when we begin to make sense of the world around us.

The Trend Management Toolkit is a system for organizing ideas and insights to fuel fresh thinking and progressive innovation strategies. When faced with a challenge or concept you want to test, you frame it by linking it to the main drivers, behaviors, and values that influence it. In practice, this means grouping and clustering information into families of ideas, data, and values by identifying the macro and micro influences.

By cataloguing data individually – collecting and eventually connecting the dots (Figure 7.1) – you get a broader perspective and recognize the interconnectedness that assists with the process of finding answers. This is the multidimensional methodology – a systemized approach to evaluating your current situation and, even more importantly, highlighting the missing elements in your business strategy. These gaps may represent potential new opportunities that could become meaningful concepts in the future. This iterative process of spotting the gaps is how we developed our trend management system in the first place. It began as a response to the need to develop a system for finding solutions to challenges not just in the present, but 3, 5, 10 or even 20 years into the future. The method has evolved over the years into the step-by-step model we use today (see Figure 7.2) and share with our clients.

Trend management is vital to corporate introspection in the 21st century. Any attempt to develop a sound future strategy requires a general theory of things or trends that factors in a broad context of influences and environments. A metaperspective allows us to understand how our findings fit into a larger ecosystem or context and it becomes invaluable when creating leadership strategies, public development programs, business concepts, or products and services.

Trend management does require us to switch from autopilot to mindful contemplation – testing each of the assumptions, systems, and processes we routinely employ without thinking – and this approach brings the rewards of clarity and strategic overview. Not only do we pinpoint where we are now, but we can start to evaluate where we want to be and how to map out the best route to that destination. Future planning is not a prescriptive process with one linear path towards the right answer, but a multilayered and collective sharing of wisdom and ideas. In other words, you cannot start by finding answers, you must start by framing the right questions.

## Unlocking the potential of trend management

The Trend Management Toolkit provides you with the ability to create a profound medium- to long-term strategy. However, in order to capitalize on the potential of these tools, you need to allocate time to think about the future you want to create, as has been explored in earlier chapters. Any forward-thinking organization integrates trend management into its business strategy.

In the case study section of this chapter, we explore how, over the years, large corporations have done just that by taking time out to think. All the organizations used the Trend Management Toolkit to navigate complexity, solve current challenges, and take control over their future. To do this, they engaged in a collaborative process that consisted of four key stages: kick-off, information, knowledge, and implementation (Figure 7.2). This approach was also iterative, meaning that the steps followed were reviewed and revisited several times during the project process in order to fine-tune the initial ideas and refine a future vision. In the following section, we describe how this step-by-step process works in practice, before moving on to the cases that illustrate the results this approach can achieve.

### Framing challenges and kick-off

**Frame:** The first step is to identify the most critical business challenges in a collaborative manner. This is likely to involve discussion of group and individual perspectives; consideration of data and market conditions; issues currently facing/expected to face the organization; external forces, including new regulations, economic developments, demographic changes; and identified disruptive influences, for instance breakthrough innovation, new business models.

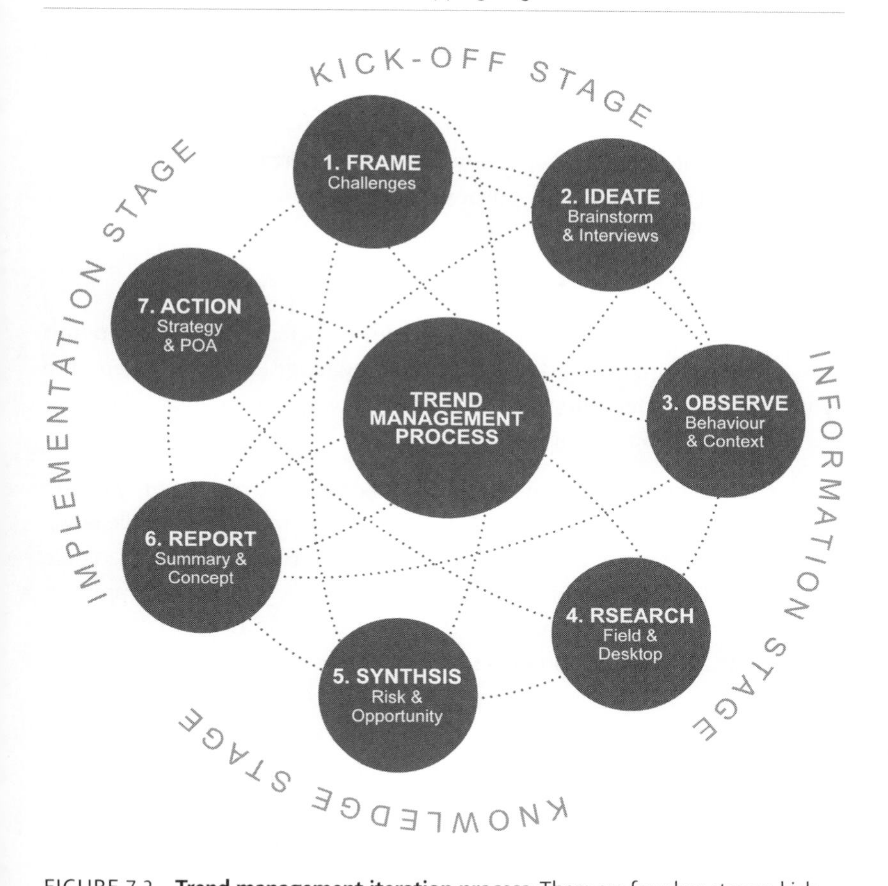


FIGURE 7.2 **Trend management iteration process**: There are four key stages: kick-off, information, knowledge, and implementation
*Source*: Kjaer Global

## Information gathering

**Ideate:** Through brainstorming sessions and expert interviews, ideas form around the core challenges and these are also the first steps in spotting opportunities that may drive strategy and innovation. It is important to consider all ideas – however radical – in order to engage in honest debate, including "what if" and "think the unthinkable" scenarios.

**Observe:** Linking challenges with core macro trends, behavior, and values informs understanding of the current situation and helps create a springboard from which meaningful concepts and solutions can develop.

**Research:** Comprehensive field and desktop research is recommended to support observation and add evidence to the decision-making process. This may be undertaken as an internal or external activity and is likely to include qualitative findings; any research undertaken should adopt a balanced 4P (people, planet, purpose, profit) approach to ensure a holistic perspective.

### Knowledge stage

**Synthesize:** After the observation and information stage, an in-depth cross-analysis is performed. The initial knowledge stage is where the synthesis is converted into concrete insights, giving a more focused picture of the risks and opportunities ahead.

**Report:** Any journey into the future requires a clear picture of the core drivers and change agents at play as the basis for formulating a viable direction. In the strategy and concept phase, report writing creates a tangible platform where the most interesting concepts are developed and supported by insights, facts, and figures. This delivers a selection of concrete recommendations and a set of potential solutions to the initial challenges.

### Trend management in action

**Implementation:** To ensure the future road map will be effectively implemented, it is essential to create a concrete plan of action (POA). It should be undertaken in collaboration with core stakeholders – to give them "ownership" – so that everyone works with the same success criteria in the implementation phase.

## Context mapping and design thinking

In the infant stage of any change process, trend management is invaluable for scanning data, collating insights, and testing initial ideas. More importantly, it helps build a framework, trial new concepts or solutions, and then test and adjust these "beta mode" ideas to get closer to the issues that really matter in an organization's value chain and people's lives. Many organizations now work with context mapping and design thinking to ensure that creative approaches are strategically incorporated into the journey from early idea conception to final product and service innovation. This process is explored by Rubino, Hazenberg, and Huiman in *Meta Products*:[3]

> in this front end, many other disciplines take place and generally these are responsible for setting up a business case: R&D departments, business advisers, market researchers, corporate strategists, etc. Design thinking seems to be the "glue" between these disciplines, because it allows the interpretation of all the transdisciplinary work into design concepts.

The discourse in the book is largely focused around how Internet-connected products and services will develop in the future. Progressive forward-thinking solutions have already entered the market, making the digital sphere the place where businesses connect to their communities and expand their value networks. But evidently, a world of Internet-connected things will, in the future, mean a mega scale "living environment," in which multidisciplinary teams collaborate to manage and navigate an interconnected reality across departments and disciplines. This will enable the development of concepts that appeal to people's diverse needs and touchpoints, as well as factoring in the business imperatives we need to thrive in the 21st century. In particular, as previously discussed, the 4P approach will become essential to any business strategy. Even when budget or resources are allocated to solve an immediate organizational challenge, this must still factor in 4P considerations to deliver medium- to long-term value.

## SUMMARY: Practical trend mapping

- It is only when we isolate and observe trends that we start spotting opportunities and make sense of the world around us.
- Trend management requires us to switch from autopilot to mindful reflection – testing systems and processes we routinely employ without thinking.
- The Trend Management Toolkit system clusters information into families of ideas, data, and values by identifying the macro and micro drivers.
- By cataloguing data individually – and eventually connecting the dots – we get a broader perspective that evaluates and highlights missing elements and potential new opportunities in business strategy.
- Future planning is not a linear path towards one right answer, but a multilayered and collective sharing of wisdom and ideas, where you start by formulating the right questions.
- Trend management is an iterative process, in which we review and revise to develop an organic working strategy, with four stages – kick-off, information, knowledge, and implementation.

## Trend management in action across sectors

This section provides illustrative cases, culled from various industries, in order to demonstrate how trend management tools can be applied across sectors in a practical manner to fulfill the diverse needs of any organization trying to solve current and potential challenges by embarking on an informed, future-oriented journey. These "live" cases – all taken from Kjaer Global's client work over the years – encompass business-to-consumer and business-to-business examples. They illustrate how trend management works in practice to solve a wide variety of business challenges, enabling the reappraisal of an organization's structure, activities, and approach in order to cultivate innovation culture and deliver meaningful solutions.

### 1. Disruptive "smart society" innovations

**Sector:** Consumer and enterprise electronics

**Challenge:** This global technology player already had a firm understanding of the importance of trend tracking, but wanted to tap into opportunities presented by the smart society and enabled by the IoT.

**Project aim:** Generate new thinking by breaking down silo culture, without losing sight of workflow continuity and pressing business objectives. This included harnessing sociocultural research to stand out as a cutting-edge connectivity brand.

#### Step-by-step process

**Framing an overarching trends ecosystem:** It was critical to the company's reputation and future success to identify tension areas and comfort zones in the integration of technologies. During research framing, specific attention was paid to intersections between "systems," such as private and public life, as key areas for disruptive innovation but also potential risk.

**Human-centric research focus:** Brainstorming and ideation sessions with the client revealed specific questions to be explored, enabling targeted desktop research within a broader multidimensional frame. Observations considered macro trend implications on a society-wide level, while micro trends explored behavior and human-scale interfaces.

**Seeing opportunity on all levels:** To pinpoint opportunities, reporting adhered to the communication parameters defined at the early ideation phase of "bigger picture" (macro) data. Results were consistently narrated in relation to people's everyday lives and backed by insights and evidence. The conclusions highlighted concrete actions for delivering true value in customer interactions, along with larger scale opportunities and disruptions of the market.

**Innovating while maintaining business flow:** The research document delivered its clear objective of enabling front-end innovation across global departments. An ecosystem of macro trends and likely scenarios tapping into micro trends was developed, with a clear distinction between impact areas – defined as personal, social, "third spaces," work – while maintaining existing company flow and objectives.

### TREND MANAGEMENT OUTCOME: Cultivating disruptive thinking

- Inspiring bold thinking on the future of the smart society and the IoT.
- Clear focus on "human-scale" risks and rewards to anchor innovation in a real-life context.
- Focus on disruptive opportunities without losing sight of core business plans.
- Encouraging an innovation culture focused firmly on people.

## 2. The humanization of financial services

**Sector:** Financial services and mortgages

**Challenge:** To become more future focused by forging new partner alliances and stronger engagement with end consumers, this key FS (financial services) player wanted to move from being a pragmatic to a more emotionally connected brand.

**Project aim:** To strengthen market position by building trust and brand awareness through an integrated communication and change management strategy. A multichannel platform for core brand messages was required to align product, service, brand, and marketing functions.

### Step-by-step process

**Online learning and co-creating:** A digital academy was created for internal stakeholders to foster co-creation and project ownership, knowledge sharing, and open dialogue. The online forum and workshops provided trend management exercises, teaching participants how to spot risk and opportunity by mapping trends, as well as facilitating a deeper understanding of how to build an emotionally engaging brand narrative.

**Making trends tangible:** Specific focus was on successfully managing stakeholder engagement through a common brand vision and perspective. To assist with this, research insights were presented in an inspiring and tangible way; tools included video case studies, infographics, and visual trend snapshots – all inviting the team to consider future customers' wants and needs.

**Vision 2020+:** The digital academy created a continuous feedback and iteration loop, encouraging the change-maker team to rethink and refine ideas over the initial six-month time frame. Additional trend workshops focused on the most innovative insights and core ideas, with the whole process documented and synthesized in a lean, practical report.

**Making the future happen:** The change-maker team embraced trend management as a co-creation process. The research findings led to a step-by-step implementation plan focusing firmly on customers and the people serving them. It inspired the team to think "from the outside in," using trend management to capitalize on continuous market disruption.

### TREND MANAGEMENT OUTCOME: Meaningful brand experiences

- A unified platform to communicate a 4P brand vision to all stakeholders.
- Practical tools to link strategy, performance, and value parameters.
- Open dialogue and co-creation to build team strengths and partnerships.
- Active future risks and opportunities management for the FS sector.
- Driving intrapreneurial culture in order to tap into disruptive FS opportunities.

## 3. Inspired innovation for a greener tomorrow

**Sector:** Automotive

**Challenge:** To drive innovation, projecting 5,10 or 15 years into the future, one of the world's largest car manufacturers wanted a macro-scale conceptual trend management process to support engineers and R&D in the early stages of planning, as well as informing its communication strategy.

**Project aim:** Develop a deep understanding of potential future automotive scenarios across the five European markets – Italy, Spain, France, Germany, and the UK. While overall focus was on future automotive scenarios, a key implicit aim was to imagine future consumer behavior and needs – envisaging how "tomorrow's people" would want to work, live, consume, and travel.

### Step-by-step process

**Developing a multidimensional system:** Identifying the key drivers of change and subsequent user segmentation of future consumers were the two core objectives and required the exploration of trends influencing society, lifestyles, value sets, and needs. The initial challenge was to develop a structure that moved beyond the organization's existing PESTEL methodology to explore interconnected macro and micro influences. A multidimensional framework was chosen as the most productive way to engage all stakeholders.

**Shaping the future building blocks:** To form a realistic basis for future ideation, backcasting was used to map past and present society events, breakthrough discoveries, and consumer trends – aligning them to automotive innovation over time. Unusually for the era, the organization also wanted the study to include what were loosely termed "speculative" elements – exploring less tangible aspects of people's value universe. This required extensive qualitative consumer research, which was built around a series of global expert interviews.

**Visualizing tomorrow's car buyers:** The six-month project collated vast amounts of diverse data before narrowing down research findings into a lean Trend Atlas verifying current and emerging lifestyle patterns and automotive needs. Having constructed core consumer profiles, the project entered the "what if?" realm: not only conceptualizing people's future values and ideals, but also imagining the automotive technology that would match them. Some data revealed weak signals of key emerging lifestyle preferences, notably in areas of intuitive smart technology, health/wellness focus, sustainability choices, and small compact solutions – all of which shaped the final four automotive scenarios.

**Incorporating a lean future vision:** This project was ahead of its time for any sector – but especially automotive – and the lifestyle scenarios presented in the final comprehensive report had a profound impact on R&D and communication strategy. The organization became a pioneer of small and sustainable automotive technology and this legacy is still reflected in its product lines and brand reputation today.

TREND MANAGEMENT OUTCOME: Connecting to tomorrow's people

- Multidimensional qualitative and quantitative data to define the direction for R&D and communication.
- A groundbreaking and holistic lifestyle study to inform future car innovation.
- Detailed profiles of tomorrow's consumer typologies and their ideal products and service experiences.
- Considering less tangible and "speculative" drivers for purchasing decisions.
- Pioneering strategies to reshape a major brand's vision for the next two decades.

## 4. Global lifestyle narratives in a local context

**Sector:** Technology and telecoms

**Challenge:** This major organization wanted to convert vast amounts of market data gathered from more than 80,000 mobile phone users globally into a lifestyle and strategic customer insight tool that would be capable of considering variations and future needs at a localized level.

**Project aim:** To develop an internal innovation platform for connecting design studios across the world and leveraging brand and marketing departments. The platform was required to enhance existing consumer segments, identify "blind spots," and inspire product and service innovation focused around local cultural needs in developed and developing markets.

### Step-by-step process

**Big data sense making:** The ideation phase centered on how best to utilize and analyze quantitative and qualitative data alongside market trends derived from a survey of 80,000+ consumers across BRIC, US, European and African markets. To make sense of the vast and diverse range of insights, the analysis of global segments was framed around core lifestyles; it focused on attitudes to technology, brands, and buying alongside values, behaviors, and involvement.

**Decoding local flavors:** Field observations, utilizing local networks and communities supported by qualitative research, added human-centric insight and cultural flavors. The synthesis phase adhered to the Trend and Lifestyle Navigator structure, with a global map of consumer profiles, clearly revealing hot spots and gaps that were potentially key areas of opportunity.

**Simplified consumer navigation:** This was a cross-divisional tool, so a major consideration was how best to organize the distilled findings into a practical brand and lifestyle handbook that would be intuitive to use. Ultimately, color coding provided the required lean visual overview, while map-based infographics with supportive presentation material presented a geographic spread of lifestyle patterns and trends across localities.

**Roll-out and ownership:** Implementation was a major challenge in such a huge company; hence, targeted lifestyle narrative workshops documenting the consumer journey were presented to internal experts across the organization in strategic local locations so key stakeholders could take "ownership." The roll-out's success was evidenced by the internal teams' continuation of research into new market opportunities – still a key part of the overall brand strategy, with notable success in emerging telecom markets.

### TREND MANAGEMENT OUTCOME: Innovation through visual storytelling

- Consolidated big data defining key global segments with regional lifestyle variants.
- "Human-centric" field and qualitative research insights.
- Focused micro level approach to expose opportunities in developed and developing markets.

- Presenting complex material in a single "visual lifestyle tool" document to bridge company departments and regions.
- Enabling "ownership" of consumer behavior trend material to foster a true culture of innovation.

---

## 5. Aligning brand experience and future strategy

**Sector:** Home decoration and DIY

**Challenge:** This home decor manufacturer with a diverse regional brand portfolio wanted a unified future-focused road map centered around its customers and their future home decor preferences.

**Project aim:** Cross-regional survey in macro trends and lifestyle preferences. The purpose was to deliver an integrated trend strategy with region-specific value for optimized product offering and engaged customer relationships.

### Step-by-step process

**Workshop-inspired concept:** The project began as a workshop survey of 200 internal marketing and sales staff across ten regions. Findings about internal stakeholders indicated such regional diversity that the company focus turned outwards – a major customer segmentation and trend study was agreed as the next logical step.

**Framing trends and consumer mindsets:** A key objective of the consumer study was to illustrate how macro trends would mature in different time frames across operational regions. Research and surveys were organized in such a way that data and findings could be examined individually and then compared to provide "big picture" analysis.

**From complex data to inspiring storytelling:** Twelve key macro trends were identified and documented in detail. The trends informed the core mindset profiles and were integrated to highlight likely consumer behavior, impact of individual trends, and potential for their growth. The segmented mindsets were narrated in depth, with visuals and graphs allowing at-a-glance comparison. Medium- to long-term opportunities and implications for each market formed the concluding section of the report.

**Towards a unified brand and new opportunities:** Since understanding the customer was evidently the key to brand strength, more integrated processes were introduced – starting with marketing and communication strategies. More resources were allocated to developing a unified global market vision, incorporating shared values and regional variations.

### TREND MANAGEMENT OUTCOME: A unified brand with a local voice

- Building understanding of consumer trends' evolution in regional contexts.
- Aligning overall brand vision with customer base and lifestyle choices.
- High-level, comprehensive consumer study to allow for easy cross-comparison by region.
- Integrating core strategy while allowing for decentralized adjustments.
- Supporting creative communities to strengthen regional core brands.

## 6. Disruptive innovation in a deal-driven market

**Sector:** Telecoms and network provider

**Challenge:** Fierce competition in a deal-driven market inspired this global telecoms provider to harness trend intelligence to find and explore new opportunities through local engagement with customers.

**Project aim:** Co-create an annual trend publication with in-house trend teams to align vision across all departments and drive innovation. The objective was to expand the client base and enhance customer retention through community engagement and CSR initiatives.

### Step-by-step process

**Sense making on a large scale:** A tightly timed schedule was agreed to handle the project so that the organization met its pressing business objectives. Existing market research was analyzed and a multidimensional platform was confirmed as the optimal route, with an initial Trend SWOT analysis required to identify further research focus.

**Framing trends in an intuitive navigator:** Global scope, with particular focus on key European and LATAM markets, was essential. This meant delivering consumer intelligence in a single Trend and Lifestyle Navigator framework, with findings measured and then benchmarked by comparison across regions. Findings were anchored by key macro trends, and supported by myriad micro level opportunities.

**Storytelling on a human scale:** Engaging local narratives was essential; they were delivered as visual trend snapshots verified by key research findings, market statistics, and infographics. Video evidence and case studies formed a core part of communicating each trend in an immersive, "human-centric" way.

**Publishing a future guide:** The final "future guide" publication package – six individual trend booklets plus a separate executive trend summary – was launched to senior management across departments in a one-day workshop, with the core findings presented through individual exercises. The publication was then disseminated by the internal global head of trends at all local offices, ensuring a shared future vision and alignment of values across this vast organization. Over several years we helped develop and expand the strategic trend tool.

### TREND MANAGEMENT OUTCOME: Market advantage through local reach

- A custom-made program to support brand and product innovation.
- Global reach, with emphasis on target European and LATAM markets.
- Inspiring local storytelling to bring customer values and needs to life.
- Strictly timed delivery and knowledge sharing, including regular updates and refresher workshops.
- Development of innovative loyalty and retention activities to strengthen market position and brand USPs.

## 7. Building extraordinary customer relationships

**Sector:** Healthcare equipment

**Challenge:** A leading specialist healthcare equipment manufacturer wanted to reinforce its positioning with a "deep insight" communication tool to understand existing customers and reach out to potential new markets.

**Project aim:** Build on existing corporate strategy work by developing a cross-functional tool for design, customer engagement, marketing, and sales. The objective was to raise the company profile as a leading developer of innovative high-end specialist equipment for medical professionals.

### Step-by-step process

**Seamless continuity and expansion:** Existing consumer profiles were revised using the mindset and persona model to align society drivers with key specialist healthcare and lifestyle trends. With the initial segmentation in place, analysis verified where previous implementation had failed, providing the groundwork for research and ideation.

**Fine-tuning audience needs:** To target the specific needs of healthcare professionals and their clients, research was focused on issues such as ergonomics, aesthetics, customer care, and financial incentives, alongside general market and lifestyle trends. These robust insights informed everything from R&D, design and color strategy through to final marketing and sales team communications.

**Four stories come to life:** Filtering of preliminary research and key macro trends led to the creation of core typologies divided into market segmentation. The next step was to bring end users' profiles to life by considering their professional mindsets and lifestyle preferences. With a 15-year plus product life cycle, research also factored in long-term drivers to project future developments.

**Durable communication platform:** The chosen communication strategy was to present research findings alongside a time line mapping company achievements and milestones since the 1950s. This made it possible to narrate stories about the brand's innovative heritage, while also delivering fresh and engaging propositions about quality healthcare technology. Five years on, the original core typologies and design strategy still inspire its innovation and communication.

---

### TREND MANAGEMENT OUTCOME: Imagining medical professional futures

---

- Healthcare-specific focus to address needs of busy professionals.
- Building an integrated innovation/communication platform considering major drivers in specialist long life cycle products.

- Developing a compelling narrative to communicate heritage and high-end specialist healthcare solutions. Projecting medical professional and end user (client) needs, and financial incentives.
- Positioning the manufacturer as a high-end innovator in its market.

## 8. Design powered by emotional benefits

**Sector:** Designer home and lifestyle products

**Challenge:** A global manufacturer of designer home and lifestyle products, with a portfolio of high-end heritage brands, wanted to support its teams and drive innovation by creating a set of cross-brand trend and lifestyle tools for product development and innovation.

**Project aim:** To enable new thinking for all brand categories, supporting design briefs and avoiding product overlap and brand diffusion. The task was to strengthen the core brand profile and drive storytelling around the twin themes of heritage and innovation.

### Step-by-step process

**Trend relevance and gap analysis:** An initial verification and gap analysis audit of existing company trend material revealed a short-term focus on design-driven micro influences only, not factoring in long-term, substantial macro shifts. This increased the risk of nonaligned products and marketing messages, hence a global scan of macro and micro trends was undertaken to support the next stage of the project.

**Communicating trend relevance:** Through cross-divisional workshops, it became evident that further insights into the interconnectedness of trends would be invaluable. A lifestyle study incorporating field and desktop research helped build and verify an exclusively developed Trend Atlas to support specific brand needs. Findings were translated into trend snapshots to support new design and product portfolio.

**Illustrating tomorrow's lifestyle behavior:** A trend and lifestyle toolbox was developed to enable product innovation to match end users' real needs and wants. Supported by a comprehensive consumer lifestyle segmentation report and executive summary, this toolkit was designed to enable knowledge sharing at all levels. Customized "nice to know" and leaner "need to know" toolboxes were created to support departmental strategy and design work across multiple brands.

**Creating future-proof concepts:** Trend and lifestyle tool implementation workshops for the group's strategic and creative teams built understanding and "future awareness" to support informed decision making. This led to a final practical consumer mindset tool, accompanied by a Trend and Lifestyle Navigator, which summarized key trends and values and included a "how to" guide to future-proofing innovation concepts. To celebrate the initial success of the consumer lifestyle study, it was verified with a Gallup market study in five regions, sampling well over 12,500 people. This project still informs the organization's future strategy work.

---

### TREND MANAGEMENT OUTCOME: Unified trend strategies

- Shifting group focus from transient micro trends to long-term macro trends.
- Developing a cross-divisional trend and lifestyle toolbox to strengthen individual brands and inspire unique product innovation.
- Creating a practical implementation platform for better decision-making processes.
- Fostering a coherent group vision by enabling the sharing of ideas.
- Strengthening overall brand profile while encouraging localized concept innovation that addressed core customer preferences.

---

## 9. A values-driven sustainability narrative

**Sector:** Retail and manufacturing

**Challenge:** This major player, already a leader in manufacturing/retail sustainability, wanted to take its powerful CSR vision to the next level by gaining further reach throughout the organization and its value chain, connecting to people in a heartfelt, authentic manner.

**Project aim:** Having developed outstanding green credentials at a strategic level, this vision needed to be redefined to increase awareness through all communication channels. The objective was to ensure values were anchored within the company culture to connect directly with all stakeholders inside and outside the organisation.

Step-by-step process

**Multidimensional thinking:** The trend gap analysis using a sustainability Trend Atlas revealed a dissonance between rational PESTEL drivers and the emotional value dimensions. While the organization outperformed competitors on quantifiable sustainability targets and activities, this advantage in performance had not been translated into engaging storytelling that resonated with people's everyday lives.

**Storytelling about the good life:** Several meetings led to the development of a storytelling framework that connected internal insights with concrete consumer trends, behavior, and core drivers from the Trend Atlas. The main driver for building a heartfelt narrative was the "good life," offering a multichannel platform to explore sustainability and tap into people's individual values.

**Meaningful brand experiences:** The documentation was a road map of future implications and opportunities observed "from the outside in." Findings were shared internally to strengthen understanding and ownership at all levels of the organization – seeking to make every stakeholder an ambassador for the new sustainability narrative.

**Championing a "people-centric" approach:** The new CSR strategy inspired the organization to focus on the purpose of its activities, by clearly linking them to people's values and needs – championing a "human-scale" storytelling approach. A public sustainability round table with cross-sector expert panels was arranged to further expand on the importance of a 4P sustainability ecosystem.

## TREND MANAGEMENT OUTCOME: Connecting via authentic storytelling

- Multidimensional platform to strengthen internal and external communication.
- Introducing 4P business model to drive an advanced future vision.
- Alignment of CSR objectives within a sustainability-focused Trend Atlas format.
- A green road map of implications and objectives using an "outside in" approach.
- Wider engagement through a human-scale sustainability narrative.

## 10. Focusing a heritage brand on the future

**Sector:** Luxury and architectural products

**Challenge:** This global heritage brand in the top-end luxury market wanted to develop a trend communication tool to support its existing range. The solution also had to be a launch pad for new product lines and themes to expand into fresh market areas – asserting its position as a partner in innovation.

**Project aim:** Foster excellent external communication touchpoints to increase market share, while also driving internal design, marketing, and communication. Cement the company's position as the "original" pioneer in its field with a longstanding heritage and innovation culture.

### Step-by-step process

**A new departure for innovation:** The launch package needed to work as an internal and external communication tool, enabling partner businesses to communicate trends direct to their customers. An audit of existing segmentation work revealed a new trend management approach was essential to create cohesive consumer universes.

**"KISS" – keep it simple sweetie:** To make the trend catalogue and launch tool accessible across sectors and languages, the design was laid out to distil insights into striking inspirational visuals. To further simplify the format, research findings were solidly anchored in quantitative and qualitative insights; real-world observations revealed new stylistic directions – first considering wider influences, then narrowcasting to core luxury market segments.

**Heritage and futurism:** The company's heritage was a key strength, but brought the threat of being perceived as too "traditional" by potential new partners. To overcome this, a luxury market Trend and Lifestyle Navigator was developed to communicate ideas rooted in heritage, as well as showcasing the wide-ranging potential for more futuristic applications to inspire new product innovation.

**Inspired collaborations:** Timed to coincide with international design fairs, the final report was so well received that the company's business partners referred to it as their "trend bible." Future design directions and customer segmentation material supported external and internal innovation and communication. The format worked wonders in redefining the brand proposition, resulting in an extensive and long-running collaboration for the trend programme across divisions.

## TREND MANAGEMENT OUTCOME: Engaging global brand experiences

- A unified product innovation and communication tool that also defined connections to customer universes.
- Simultaneously celebrating heritage and suggesting innovative new applications.
- Inspired storytelling using visual scenarios focused on style, interior, and architectural sectors.
- Efficient and engaged communication with existing and potential partners.
- Redefining the brand to cement its market position and drive growth.

## SUMMARY: Trends in action across sectors in context

### The scientific dimension

**TECHNOLOGY FOCUS**

**1. Disruptive "smart society" innovations**

**Market:** Consumer and enterprise electronics

**Objective:** Harnessing sociocultural research for innovation focused around "smart society" and IoT, also breaking down silo culture.

**Outcome:** Research document enabling front-end innovation; clear outline of potential risks/rewards in key areas of personal, social, "third space," and working environments; inspiring cross-company innovation.

**ECONOMICS FOCUS**

**2. The humanization of financial services**

**Market:** Financial services and mortgages

**Objective:** Forge new partnerships and stronger relationships with end consumers by adopting an emotionally engaging brand strategy.

**Outcome:** Unified trend road map to communicate a 4P brand vision; practical tools to enable FS product and service design with clearly defined customer value in a market ripe for disruptive innovation.

**ENVIRONMENT AND LEGISLATION FOCUS**

**3. Inspired innovation for a greener tomorrow**

**Market:** Automotive

**Objective:** Conceptual global trend scanning with European focus to inspire profound innovation matching people's lifestyles and values.

**Outcome:** Pioneering study collating vast amounts of data into a unified platform to shape R&D and brand communication – leading to the development of next-generation cars for "tomorrow's people."

## The social dimension

**COMMUNICATIONS FOCUS**

**4. Global lifestyle narratives in a local context**

**Market:** Technology and telecoms

**Objective:** Converting vast amounts of global market data into a lean customer insight tool to develop localized mobile phone products and services.

**Outcome:** A visual lifestyle tool connecting global design, brand, and marketing departments to drive innovation and highlight specific opportunities locally in developed and developing markets.

**SOCIAL STRUCTURES FOCUS**

**5. Aligning brand experience and future strategy**

**Market:** Home decoration and DIY

**Objective:** Develop a unified future-focused road map recognizing regional variations to enable strategic decisions at local level.

**Outcome:** Alignment of overall brand vision and values through consumer lifestyle study that projected trend evolution across markets and improved customer engagement.

**ORGANIZATIONAL FOCUS**

**6. Disruptive innovation in a deal-driven market**

**Market:** Telecoms and network provider

**Objective:** Harness trend intelligence to spot new disruptive opportunities in localized close engagement with people.

**Outcome:** Global trend publication aligned vision across departments and used local narratives to build customer loyalty and retention programs, particularly in target European and LATAM markets.

## The emotional dimension

### HEALTH AND WELLBEING FOCUS

#### 7. Building extraordinary customer relationships

**Market:** Healthcare equipment

**Objective:** Reinforce positioning as professional-focused innovator with a tool to support all initiatives, from R&D to communication.

**Outcome:** Comprehensive and durable lifestyle research platform focused around needs of medical professionals, factoring in major drivers impacting long life cycle health equipment.

### LIFESTYLE CHOICES FOCUS

#### 8. Design powered by emotional benefits

**Market:** Designer home and lifestyle products

**Objective:** Tools to support cross-divisional innovation and strengthen positioning of a diverse portfolio of heritage brands.

**Outcome:** Universal trend management toolbox shifted focus from transient micro trends to long-term drivers; this strengthened the individual brands, enabled idea sharing, and encouraged local concept innovation and a unique approach to design.

### A BETTER WORLD FOCUS

#### 9. A values-driven sustainability narrative

**Market:** Retail and manufacturing

**Objective:** Promote and cement recognized leadership in sustainability by communicating through "human-centric" authentic storytelling.

**Outcome:** CSR activities were aligned with 4P objectives using the multidimensional platform – introducing engaging sustainability narratives to engage and inspire all stakeholders.

---

## The spiritual dimension

### QUALITY OF LIFE FOCUS

#### 10. Focusing a heritage brand on the future

**Market:** Luxury and architectural products

**Objective:** Trend communication tool to support existing range, while building new market share through product innovation.

**Outcome:** Simultaneous celebration of heritage and new innovative design possibilities helped to cement market position, also engaging potential new partners in luxury markets.

---

## Summary: Shaping tomorrow today

In 1992, while working in my design studio in East London, my mind started to wander; I began to contemplate what I really wanted for the future of my company, where I saw myself in 10 years' time, and how to make it happen. A few weeks later, while visiting a William Blake exhibition at Tate Britain, I spotted the answer right there in front of me – it was, quite literally, the writing on the wall: "I must create a system or be enslav'd by another man's." Blake's words inspired me to question the path I was on; they reminded me that I could start actively shaping a whole new vision based around an alternative approach to understanding our world and doing business – one that factored in people-centric values, sustainable concepts, and inspiring storytelling. In short, I decided to create the future.

Blake's words resonated deeply with me then and still do today, because they address the fundamental need to shape the future, and the importance of creating our own system for making sense of the world. The trend management method presented in this book is a practical approach we trust in implicitly, but it is not prescriptive; everyone we work with embarks on a unique journey in which we guide them to take control and ownership over their future. The same will be true of your journey when you use the framework and tools provided in *The Trend Management Toolkit*.

### Investment and reward

In undertaking an assessment of our future business vision and strategy, we have to be honest with ourselves and have a genuine willingness to face tough questions in a spirit of openness, whether it is exploring past failures or recognizing where a current approach may fall short in delivering true benefits to those inside and outside our organization. Developing future concepts that offer meaning and value – at every level – is key to success in the 21st century; as many business sectors have already learned the hard way, basing any kind of strategy on a pure profit principle is likely to undermine your organization's trust ratings and, ultimately, competiveness. This is why the 4P principle of people, planet, purpose, and then profit is such a crucial frame of reference and checklist. If you don't factor all four P elements into your future vision, then it's unlikely to sustain you in tomorrow's business world.

The approaches set out in this practical guide to the future are designed to inspire visionary and fruitful strategies, but the business case for incorporating trend management within your organization is clear; the value starts immediately. In a world so focused on the "now," taking time out to think about the future is hugely rewarding and crucial to producing long-lasting results. So often, the process of sitting around a table to discuss tomorrow's challenges in a focused dialogue unlocks our ability to consider the "outside in" and commonsense perspectives that also transform current business processes and practices. The journey of future-proofing our business starts when we honestly address two fundamental questions: Where are we now? Where do we want to be?

Mapping out where you want your business to be in 5, 10 or 20 years' time is not simple or immediate – there is no one "right" answer, but many possible answers; however, in asking the right questions, you take the first vital step to creating a viable future strategy. The Trend Atlas you design is your unique road map, a flexible framework that can be reviewed and updated as new trends emerge and market conditions shift. Creating the initial road map takes effort, but the rewards – not least revisiting the purpose of your business operation – repay the initial investment.

It is liberating to remind ourselves that the future is not some place we go to, but something we create; it reminds us that we all have the capacity to be active change-makers and trend influencers and, ultimately, visionary leaders shaping better organizations. This is surely the most compelling argument for incorporating trend management into your daily practice and planning. In this spirit of endless possibilities and discoveries on the road ahead, I wish you a fruitful journey to creating your unique future business vision.

# NOTES

## Chapter 1: From facts to feelings


1. Adams T. This much I know: Daniel Kahneman. *The Observer.* 8 July 2012. www.theguardian.com/science/2012/jul/08/this-much-i-know-daniel-kahneman (accessed 7 November 2013).
2. Hetman F. *Le Langage de la Prévision* (*The Language of Forecasting*). Paris. Futuribles/S.E.D.E.I.S. 1969.
3. Goetz K. How 3M Gave Everyone Days Off and Created an Innovation Dynamo. fastcodesign.com. 1 February 2011. www.fastcodesign.com/1663137/how-3m-gave-everyone-days-off-and-created-an-innovation-dynamo (accessed 13 July 2014).
4. Taleb N.N. *Fooled by Randomness: The Hidden Role of Chance in Life and in the Markets*. London. Texere. 2001.
5. Taleb N.N. *The Black Swan: The Impact of the Highly Improbable.* London. Penguin. 2008.
6. Popper R. *Mapping Foresight: Revealing how Europe and other World Regions Navigate into the Future*. European Foresight Monitoring Network. European Union. 2009. http://ec.europa.eu/research/social-sciences/pdf/efmn-mapping-foresight_en.pdf (accessed 1 November 2013).
7. Copenhagen Institute for Futures Studies (CIFS). Meaningful Consumption. www.cifs.dk/time2think/about.asp (accessed 8 November 2013).
8. Jensen R. *The Dream Society: How the Coming Shift from Information to Imagination Will Transform Your Business.* New York. McGraw-Hill. 2001.
9. De Bono E. *The Use of Lateral Thinking.* London. Jonathan Cape. 1967.
10. Hindle T. *Guide to Management Ideas and Gurus.* London. Economist Books. 2012, p. 318.
11. Wack P. Scenarios: Uncharted Waters Ahead. *Harvard Business Review,* 63(5): 72–9. 1985. http://hbr.org/1985/09/scenarios-uncharted-waters-ahead/ar/1 (accessed 6 November 2013).
12. Senge P. *The Fifth Discipline: The Art and Practice of the Learning Organization.* New York. Doubleday. 1990.

13 De Geus A. *The Living Company: Growth, Learning and Longevity in Business*. Boston. Harvard Business School Press. 1997.

## Chapter 2: Sense making in a fast-forward society

1 Ravallion M., Chen S. and Sangraula P. *Dollar a Day Revisited*. The World Bank Development Research Group, Policy Research Working Paper 4620. 2008.
2 Stiglitz J.E., Sen A. and Fitoussi J.P. *Report by the Commission on the Measurement of Economic Performance and Social Progress*. CMEPSP. 2009, p. 9. www.stiglitz-sen-fitoussi.fr/documents/rapport_anglais.pdf (accessed 3 November 2013).
3 *The Pyramid of Waste – The Lightbulb Conspiracy* (documentary originally aired on Norwegian channel NRK). 2010. https://archive.org/details/PlannedObsolescenceDocumentary (accessed 19 February 2013).
4 Mackey J. and Sisodia R. *Conscious Capitalism: Liberating the Heroic Spirit of Business*. Boston. Harvard Business Press. 2013, p. 76.
5 Klein N. *No Logo*. Toronto. Knopf Canada. 1999.
6 Singer T. Beyond Homo Economicus. Project Syndicate. 2013. www.project-syndicate.org/commentary/a-new-model-of-human-behavior-by-tania-singer (accessed 9 November 2013).
7 Jobs S. *2005 Stanford Commencement Address*. www.youtube.com/watch?v=UF8uR6Z6KLc (accessed 1 November 2013).
8 Maslow A.H. A Theory of Human Motivation. *Psychological Review*, 50(4): 370–96. 1943.
9 Veblen T. *The Theory of the Leisure Class: An Economic Study of Institutions*. Oxford. OUP. [1899] 2009.
10 Curtis A. *The Century of the Self*. BBC Four. 2002.
11 Wilson J. *Book Review: The Economics of Enough: How to Run the Economy as if the Future Matters*. March 2011. eprints.lse.ac.uk/38060/1/blogs.lse.ac.uk-Book_Review_The_Economics_of_Enough_How_to_Run_the_Economy_as_if_the_Future_Matters.pdf.
12 Keynes J.M. Economic Possibilities for our Grandchildren, in *Essays in Persuasion*. New York: W.W. Norton & Co. [1930] 1963, pp. 358–73.
13 Roenneberg T. *Internal Time: Chronotypes, Social Jet Lag, and Why You're So Tired*. London. Harvard University Press. 2012.
14 Franzen, J. Jonathan Franzen: what's wrong with the modern world. *The Guardian*. 13 September 2013.
15 Coote A., Franklin J. and Simms A. *21 Hours: Why a Shorter Working Week can Help us all to Flourish in the 21st Century*. New Economics Foundation. 2010, p. 2. http://s.bsd.net/nefoundation/default/page/-/files/21_Hours.pdf (accessed 12 November 2013).

16 London B. *Ending the Depression Through Planned Obsolescence.*1932. London_(1932)_Ending_the_depression_through_planned_obsolescence.pdf.
17 *The Pyramid of Waste – The Lightbulb Conspiracy* (documentary originally aired on Norwegian channel NRK). 2010. https://archive.org/details/PlannedObsolescenceDocumentary (accessed 19 February 2013).
18 Packard V. *The Waste Makers.* New York. Ig Publishing. [1960] 2011.
19 Consumer Federation of America. www.consumerfed.org.
20 Baudrillard J. The System of Objects, in *Jean Baudrillard: Selected Writings*, ed. and Intro. M. Poster. Stanford University Press. [1968] 2001, p. 15.
21 Botsman R. and Rogers R. *What's Mine is Yours: The Rise of Collaborative Consumption.* New York. Harper Business. 2010.
22 Ellen MacArthur Foundation. *What is the Circular Economy 100?* www.ellenmacarthurfoundation.org/business/ce100 (accessed 6 November 2013).
23 PwC. *CDP Sector Insights: What is Driving Climate Change Action in the World's Largest Companies? Global 500 Climate Change Report 2013.* www.cdproject.net/CDPResults/CDP-Global-500-Climate-Change-Report-2013.pdf (accessed 11 November 2013).
24 Myers D. The Secret to Happiness. *Yes!* magazine. 18 June 2004. www.yesmagazine.org/issues/what-is-the-good-life/866 (accessed 19 February 2013).
25 Baudrillard J. *The Consumer Society: Myths and Structures*. London. Sage. [1970] 1998, p. 193.
26 Skidelsky R. and Skidelsky E. *How Much is Enough? The Love of Money, and the Case for the Good Life*. London. Other Press. 2012.
27 *The Economist*. The Big Rethink. www.economistinsights.com/technology-innovation/event/big-rethink-2012 (accessed 11 November 2013).
28 Kjaer A.L. Emotional Consumption. Copenhagen Institute for Futures Studies. Future Orientation. February 2006. www.iff.dk/scripts/artikel.asp?id=1364&lng=2 (accessed 19 February 2013).
29 Keynes M. *The General Theory of Employment, Interest and Money.* Basingstoke. Palgrave Macmillan. [1936] 2007, Preface.
30 *The Economist.* The Nordic Countries: The Next Supermodel. 2 February 2013. www.economist.com/news/leaders/21571136-politicians-both-right-and-left-could-learn-nordic-countries-next-supermodel (accessed 19 October 2013).
31 Havas Worldwide. *Prosumer Report: This Digital Life*, vol. 13, pp. 3, 31. 2012. www.prosumer-report.com/blog/wp-content/uploads/downloads/2012/09/this-digital-life.pdf (accessed 10 November 2013).
32 Haque U. *Betterness: Economics for Humans*. London. Harvard Business Press Books. 2011.
33 Haque U. *The New Capitalist Manifesto: Building a Disruptively Better Business.* London. Harvard Business Press Books. 2011.

34 Kruger B. *Untitled (I Shop Therefore I Am)*. MoMa Collection. New York. 1987.
35 Martineau P. *Motivation in Advertising: The Motives that Make People Buy.* New York. McGraw-Hill. 1957, p. 73
36 Havas Media. Meaningful Brands. 2013. www.havasmedia.com/meaningful-brands (accessed 10 November 2013).
37 Fisk P. *People, Planet, Profit: How to Embrace Sustainability for Innovation and Business Growth.* London. Kogan Page. 2010.

## Chapter 3: Trend mapping: past, present and future

1 Poincaré H. *Science and Method*, trans. Francis Maitland. Mineola, NY. Dover. [1914] 2003, p. 129.
2 Georghiu L., Cassingena Harper J., Keenan M., Miles I. and Popper R. (eds). *The Handbook of Technology Foresight: Concepts and Practice*. Cheltenham. Edward Elgar. 2008.
3 Curtis A. *The Century of the Self*. BBC Four. 2002.
4 Marcuse, H. Interview 1967, Curtis A. *The Century of Self*. BBC Four. 2002.
5 Kahneman D. The Riddle of Experience vs. Memory. TED. February 2010. www.youtube.com/watch?v=XgRlrBl-7Yg (accessed 1 December 2013).
6 Gopnik A. Mindless: The New Neuro-skeptics. *The New Yorker*. 9 September 2013. www.newyorker.com/arts/critics/books/2013/09/09/130909crbo_books_gopnik (accessed 8 December 2013).
7 Kahneman D. *Thinking, Fast and Slow.* London. Penguin. 2011.
8 McGilchrist I. *The Master and his Emissary: The Divided Brain and the Making of the Western World.* London. Yale University Press. 2009.
9 Nielson J. Researchers Debunk Myth of "Right-brain" and "Left-brain" Personality Traits. University of Utah Health Care. 14 August 2013. http://healthcare.utah.edu/publicaffairs/news/current/08-14-2013_brain_personality_traits.php (accessed 10 July 2014).
10 Kandel E.R. The New Science of Mind. *The New York Times*. 6 September 2013. www.nytimes.com/2013/09/08/opinion/sunday/the-new-science-of-mind.html?pagewanted=all&_r=0 (accessed 28 April 2014).
11 OECD. *Understanding the Brain: Towards a New Learning Science*. OECD Publishing. 2002. www.oecd.org/edu/ceri/31706603.pdf (accessed 8 December 2013).
12 BBC News. Brain changes seen in cabbies who take The Knowledge. 8 December 2011. www.bbc.co.uk/news/health-16086233 (accessed 8 December 2013).
13 Thompson C. *Smarter Than You Think: How Technology is Changing Our Minds for the Better.* New York. Penguin Press. 2013, p. 284.

14 Wegner D.M. *Transactive Memory: A Contemporary Analysis of the Group Mind*. New York. Springer. 1986.

15 Geist W.E. Pondering the year 2000 with Kahn, Haig & Co. *The New York Times*. October 1982. www.nytimes.com/1982/10/02/nyregion/pondering-the-year-2000-with-kahn-haig-co.html (accessed 28 April 2014).

16 Clark R.C. and Mayer R.E. *E-Learning and the Science of Instruction*. San Francisco. Pfeiffer. 2007.

17 Sontag S. *On Photography*. New York. Picador USA. 1977, p. 3.

18 LSE Public Policy Group. *Maximizing the Impacts of your Research: A Handbook for Social Scientists*. 2011. www.lse.ac.uk/government/research/resgroups/lsepublicpolicy/docs/lse_impact_handbook_april_2011.pdf (accessed 28 April 2014).

19 Kurzweil R. and Diamandi P. Singularity University. http://singularityu.org.

20 Miller R. 21st Century Transitions: Opportunities, Risks and Strategies for Governments and Schools. In *What Schools for the Future?* OECD. 2001. www.oecd.org/site/schoolingfortomorrowknowledgebase/futuresthinking/trends/41286179.pdf (accessed 10 December 2013).


## Chapter 4: Your essential trend toolkit


1 Beck U., Giddens A. and Lash S. *Reflexive Modernization: Politics, Tradition and Aesthetics in the Modern Social Order*. California. Stanford University Press. 1994, p. 177.

2 Foucault M. The *Order of Things: An Archeology of the Human Sciences*. London. Routledge. 1989.

3 Kao J. *Jamming: The Art and Discipline of Business Creativity*, 2nd edn. New York. HarperCollins. 1996.


## Chapter 5: Major trends to 2030+


1 *Millennials at Work: Reshaping the Workplace*. London. PwC UK. 2011, p. 3. www.pwc.com/en_M1/m1/services/consulting/documents/millennials-at-work.pdf (accessed 12 February 2014).

2 *Dementia: A Public Health Priority*. Geneva. WHO. 2012, p. 6. http://apps.who.int/iris/bitstream/10665/75263/1/9789241564458_eng.pdf (accessed 12 February 2014).

3 *IDF Diabetes Atlas*, 6th edn. Brussels. International Diabetes Federation. 2013, pp. 7, 11. www.idf.org/diabetesatlas (accessed 12 February 2014).

4 Finkelstein E.A. et al. (2012) Obesity and Severe Obesity Forecasts Through 2030. *American Journal of Preventive Medicine.* 42(6): 563–70, 2012. www.ajpmonline.org/article/S0749-3797(12)00146-8/fulltext (accessed 12 February 2014).

5 *World Urbanization Prospects: The 2011 Revision*. New York. UN Department of Economic & Social Affairs. 2011. http://esa.un.org/unup/pdf/FINAL-FINAL_REPORT%20WUP2011_Annextables_01Aug2012_Final.pdf (accessed 12 February 2014).

6 Gupta M. Internet's governance can't be limited to one geography. *The Times of India.* 30 July 2012. http://articles.timesofindia.indiatimes.com/2012-07-30/edit-page/32924041_1_internet-governance-internet-corporation-root-servers (accessed 13 February 2014).

7 Gnanasambandam C., Madgavkar A., Kaka N., Manyika J., Chui M., Bughin J. and Gomes M. *Online and Upcoming: The Internet's Impact on India.* Bangalore. McKinsey & Company. 2012, p. 3. www.mckinsey.com/~/media/McKinsey%20Offices/India/PDFs/Online_and_Upcoming_The_internets_impact_on_India.ashx (accessed 13 February 2014).

8 *2012 Edelman Trust Barometer*. Edelman. 2012. www.edelman.com/insights/intellectual-property/2012-edelman-trust-barometer. (accessed 13 February 2014).

9 *Connections Counter: The Internet of Everything in Motion*. The Network. Cisco. 2013. http://newsroom.cisco.com/feature-content?type=webcontent&articleId=1208342 (accessed 10 February 2014).

10 Rooney J. Havas CEO David Jones on Why Good Business Is No Longer Optional. *Forbes*. 28 February 2012. www.forbes.com/sites/jenniferrooney/2012/02/28/havas-ceo-david-jones-on-why-good-business-is-no-longer-optional/ (accessed 12 February 2014).

11 Faulk R. Why Total Transparency is Good for Business. Openview Labs. 18 October 2013. http://labs.openviewpartners.com/why-corporate-transparency-is-good-for-business. (accessed 14 February 2014).

12 The secret of their success. *The Economist*. 2 February 2013. www.economist.com/news/special-report/21570835-nordic-countries-are-probably-best-governed-world-secret-their (accessed 14 February 2014).

13 *The Nordic Way: Shared Norms for the New Reality.* Geneva. World Economic Forum and Global Challenge. 2011, p. 27. www.globalutmaning.se/wp-content/uploads/2011/01/Davos-The-nordic-way-final.pdf (accessed 13 January 2014).

14 *Kelly Global Workforce Index. Effective Employers: The Evolving Workforce.* Kelly Services. 2011, p. 14. www.kellyocg.com/Knowledge/Kelly_Global_Workforce_Index_Content/Effective_Employers_Report/ (accessed 13 January 2014).

15 Clark L. Sweden's early adopter foreign minister on crafting digital diplomacy. *Wired*. 17 January 2014. www.wired.co.uk/news/archive/2014-01/17/carl-bildt-digital-diplomat (accessed 13 January 2014).
16 Randers J. *2052: A Global Forecast for the Next Forty Years*. Vermont. Chelsea Green Publishing. 2012, p. 15.
17 Birch S. Open thread: co-operatives – what's in store for 2014? *The Guardian*. 17 December 2013. www.theguardian.com/social-enterprise-network/2013/dec/17/co-ops-social-enterprise-2014 (accessed 13 January 2014).
18 Dezenski E. *The Business Case for Fighting Corruption*. Transparency International. 24 September 2012. http://blog.transparency.org/2012/09/24/the-business-case-for-fighting-corruption/ (accessed 9 January 2014).
19 Faulk R. Why Total Transparency is Good for Business. Openview Labs. 18 October 2013. http://labs.openviewpartners.com/why-corporate-transparency-is-good-for-business (accessed 14 February 2014).
20 Hall J. 10 Leaders Who Aren't Afraid To Be Transparent. *Forbes*. 27 August 2012. www.forbes.com/sites/johnhall/2012/08/27/10-leaders-who-arent-afraid-to-be-transparent (accessed 14 February 2014).
21 *Global Health Observatory*: *Urban Population Growth*. WHO. 2014. www.who.int/gho/urban_health/situation_trends/urban_population_growth_text/en (accessed 14 February 2014).
22 Dobbs R., Remes J. and Smit S. Urban economic clout moves east. *McKinsey Quarterly*. March 2011. www.mckinsey.com/insights/economic_studies/urban_economic_clout_moves_east (accessed 14 February 2014).
23 King R. GE hopes the industrial internet will mean the end of downtime. *The Wall Street Journal*. 26 November 2012. http://blogs.wsj.com/cio/2012/11/26/ge-hopes-the-industrial-internet-will-mean-the-end-of-downtime (accessed 14 February 2014).
24 Feuer A. The mayor's geek squad. *The New York Times*. 23 March 2013. www.nytimes.com/2013/03/24/nyregion/mayor-bloombergs-geek-squad.html?pagewanted=all&_r=0 (accessed 14 February 2014).
25 Gangadharan S.P. How can big data be used for social good? *The Guardian*. 30 May 2013. www.theguardian.com/sustainable-business/how-can-big-data-social-good (accessed 14 February 2014).
26 *Ushahidi: Crowd-sourced Maps that Advance Human Rights*. MacArthur Foundation, Macarthur Awards. 27 February 2013. www.macfound.org/maceirecipients/66/ (accessed 1 March 2013).
27 *Climate Change*. UN-Habitat. http://unhabitat.org/urban-themes-2/climate-change/ (accessed 29 April 2014).
28 *Global Innovators: International Case Studies on Smart Cities*. London. Department for Business, Innovation and Skills. Paper no. 135 (BIS/13/1216). October 2013, p. 2. www.gov.uk/government/uploads/system/uploads/

attachment_data/file/249397/bis-13-1216-global-innovators-international-smart-cities.pdf (accessed 14 February 2014).

29 National Ecological Observatory Network (NEON). www.neoninc.org (accessed 14 February 2014).

30 Lohr S. McKinsey: The $33 trillion technology payoff. *The New York Times*. 22 May 2013. http://bits.blogs.nytimes.com/2013/05/22/mckinsey-the-33-trillion-technology-payoff/?_php=true&_type=blogs&_r=0› (accessed 14 February 2014).

31 *Engaging Tomorrow's Consumer*. World Economic Forum. January 2013, p. 5. www3.weforum.org/docs/WEF_RC_EngagingTomorrowsConsumer_Report_2013.pdf (accessed 14 February 2014).

32 Kiron D. et al. Introduction: sustainability, innovation and profits. *MIT Sloan Management Review*. February 2013. http://sloanreview.mit.edu/reports/sustainability-innovation/introduction/ (accessed 13 February 2014).

33 Blake A. China commits to circular economy. *Resource*. 15 February 2013. www.resource.uk.com/article/News/China_commits_circular_economy-2757#.Uv5WWv2AbEw (accessed 14 February 2014).

34 *Towards the Circular Economy: Accelerating the Scale-up across Global Supply Chains*. World Economic Forum. January 2014, p. 3. www3.weforum.org/docs/WEF_ENV_TowardsCircularEconomy_Report_2014.pdf (accessed 14 February 2014).

35 *The Global Cleantech Innovation Index Report 2012*. Cleantech Group and WWF. 2012, p. 15. http://info.cleantech.com/2012InnovationIndex.html (accessed 12 February 2014).

36 *Clean Energy Trends 2009*. Clean Edge. 2009. https://cleanedge.com/reports/Clean-Energy-Trends-2009 (accessed 14 February 2014).

37 Cheung A. and Lawn M. *Executive Summary*. Bloomberg New Energy Finance Smart Power Leadership Forum. Copenhagen. 7–8 November 2013. http://bnef.folioshack.com/document/smartpower/1c7loo (accessed 14 February 2014).

38 *2050 Something's Green in the State of Denmark: Scenarios for a Sustainable Economy*. Mondaymorning and Realdania. 2012, pp. 60, 79. www.sustainia.me/resources/publications/mm/2050.pdf (accessed 14 February 2014).

39 Youngman R. and Parad. M. *2013 Global Cleantech 100*. Cleantech Group. 2013, p. 26. www.cleantech.com/wp-content/uploads/2013/10/GCT100_Report_Digital_FINAL.pdf (accessed 13 February 2014).

40 *Towards the Circular Economy*. The Ellen MacArthur Foundation. vol 2. 2013, p. 5. www.ellenmacarthurfoundation.org/circular-economy/circular-economy/new-report-towards-the-circular-economy-vol-2 (accessed 14 February 2014).

41 Kongshøj Madsen, P. *EEO Review: Promoting Green Jobs throughout the Crisis*. Aalborg: European Employment Observatory. 2013, p. 6. www.eu-employment-observatory.net/resources/reviews/Denmark-EEO-GJH-2013.pdf (accessed 9 June 2014).

42 Nichols W. Global energy can be 95 per cent renewable by 2050, says WWF. *Business Green*. 3 February 2011. www.businessgreen.com/bg/news/2023956/global-energy-cent-renewable-2050-wwf. Full report: Singer S. et al. (ed.) *The Energy Report: 100% Renewable Energy by 2050*. WWF International. 2011. www.wwf.or.jp/activities/lib/pdf_climate/green-energy/WWF_EnergyVisionReport.pdf (accessed 14 February 2014).

43 *Build It Back Green for New Orleans*. Global Green USA. 2013. www.globalgreen.org/articles/global/75 (accessed 14 February 2014).

44 *The Future Report 2012*. Steria and Global Future and Foresight. 2012, p. 21. www.steria.com/fileadmin/com/sharingOurViews/publications/files/Steria_Future_Report_2012.pdf (accessed 12 February 2014).

45 Yueh L. The rise of the global middle class. BBC News. 29 June 2013. www.bbc.co.uk/news/business-22956470 (accessed 14 February 2014).

46 *The World In 2025: Rising Asia and Socio-ecological Transition*. European Commission. 2009, p. 9. http://ec.europa.eu/research/social-sciences/pdf/the-world-in-2025-report_en.pdf (accessed 14 February 2014).

47 *Global Trends 2030: Alternative Worlds*. National Intelligence Council. December 2012, p. 16. http://globaltrends2030.files.wordpress.com/2012/11/global-trends-2030-november2012.pdf (accessed 14 February 2014).

48 *Trend Compendium 2030*. Roland Berger Strategy Consultants. 2011, pp. 22, 35, 42, 43, 103, 121. www.rolandberger.com/gallery/trend-compendium/tc2030/content/assets/trendcompendium2030.pdf (accessed 14 February 2014).

49 *Global Europe 2050*. European Commission. 2012, p. 15. http://ec.europa.eu/research/social-sciences/pdf/global-europe-2050-report_en.pdf (accessed 14 February 2014).

50 China Dream. JUCCE. 2014. http://juccce.org/chinadream (accessed 14 February 2014).

51 O'Neill J. Who You Calling a BRIC? *Bloomberg Opinion*. 12 November 2013. www.bloomberg.com/news/2013-11-12/who-you-calling-a-bric-.html (accessed 14 February 2014).

52 Smith L.C. *The World in 2050: Four Forces Shaping Civilization's Northern Future*. New York. Plume. 2011, p. 7.

53 Dobbs R. et al. *Urban World: Cities and the Rise of the Consuming Class*. McKinsey Global Institute. June 2012, pp. 24, 36. www.mckinsey.com/insights/urbanization/urban_world_cities_and_the_rise_of_the_consuming_class (accessed 14 February 2014).

54 India's growing middle class. *Livemint & Wall Street Journal*. www.livemint.com/Opinion/jOd9cKy4pQoN91iZccnUWP/Indias-growing-middle-class.html (accessed 12 February 2014).

55 Magni M. and Poh F. Winning the battle for China's new middle class. *McKinsey Quarterly*. June 2013. www.mckinsey.com/insights/consumer_and_retail/winning_the_battle_for_chinas_new_middle_class (accessed 14 February 2014).

56 Clancy R. Asia 'to take over half luxury goods market'. *Daily Telegraph*. 12 August 2013. www.telegraph.co.uk/finance/newsbysector/retailandconsumer/10236298/Asia-to-take-over-half-luxury-goods-market.html (accessed 14 February 2014).
57 Mukherjee Parikh R. Indians, Chinese see luxury goods differently: Study. *The Times of India*. 24 November 2013. http://articles.timesofindia.indiatimes.com/2013-11-24/ahmedabad/44411694_1_luxury-brands-assocham-yes-bank-china-and-india (accessed 14 February 2014).
58 Kelion L. CES 2014: Samsung Smart Home aims to connect devices. BBC News. 5 January 2014. www.bbc.co.uk/news/technology-25616575 (accessed 14 February 2014).
59 Mee-yoo K. Four silver towns for old to be built. *The Korea Times*. 4 November 2009. www.koreatimes.co.kr/www/news/nation/2014/02/113_54888.html (accessed 17 February 2014).
60 *Stockholm Royal Seaport – a matter of attraction*. EBR. 2. 2011, p. 35. www.ericsson.com/res/thecompany/docs/publications/business-review/2011/issue2/stockholm-royal.pdf (accessed 15 February 2014).
61 Cohen B. The 10 Smartest Cities on the Planet. *FastCompany*. 11 January 2012. www.fastcoexist.com/1679127/the-top-10-smart-cities-on-the-planet (accessed 16 February 2014).
62 *Smart Cities will be enabled by Smart IT*. Steria. 2011, p. 7. www.steria.com/fileadmin/com/sharingOurViews/publications/files/STE3899-Smart_Cities_brochure_08_APP.PDF (accessed 16 February 2014).
63 *Global Europe 2050*. European Union. 2012, p. 18. http://ec.europa.eu/research/social-sciences/pdf/global-europe-2050-report_en.pdf (accessed 17 February 2014).
64 *The Societal Impact of the Internet of Things*. BCS-OII Forum Report. 2013, p. 7. www.bcs.org/upload/pdf/societal-impact-report-feb13.pdf (accessed 17 February 2014).
65 Honigsbaum M. Meet the new generation of robots. They're almost human… *The Observer*. 15 September 2013. www.theguardian.com/technology/2013/sep/15/robot-almost-human-icub (accessed 17 February 2014).
66 *Wearable Computing Devices, Like Apple's iWatch, Will Exceed 485 Million Annual Shipments by 2018*. ABI Research. 21 February 2013. www.abiresearch.com/press/wearable-computing-devices-like-apples-iwatch-will (accessed 16 February 2014).
67 Wilson H.J. Six Numbers Reveal the Booming Business of Auto-Analytics. *Harvard Business Review*. 16 May 2013. http://blogs.hbr.org/2013/05/six-numbers-reveal-the-booming (accessed 17 February 2014).
68 *Talent Mobility 2020: The Next Generation of International Assignments*. PwC. 2010, p. 6. www.pwc.com/gx/en/managing-tomorrows-people/future-of-work/pdf/talent-mobility-2020.pdf (accessed 13 February 2014).

69 *How Millennials are Changing the Way we Work*. Young Entrepreneur Council. 26 June 2012. http://theyec.org/how-millennials-are-changing-the-way-we-work (accessed 17 February 2014).

70 *Encouraging Women Entrepreneurs*. European Commission. 2013. http://ec.europa.eu/enterprise/policies/sme/promoting-entrepreneurship/women/index_en.htm (accessed 17 February 2014).

71 Maitland A. and Thomson P. *Future Work: How Businesses Can Adapt and Thrive in the New World of Work*. Basingstoke. Palgrave Macmillan. 2011.

72 Meilach D. 5 Trends Entrepreneurs Will See in 2013. *Business News Daily*. 4 January 2013. www.businessnewsdaily.com/3688-entrepreneur-trends-new-year.html (accessed 17 February 2014).

73 Empowering Women – and Men. *Business Today*. 7 July 2013. http://businesstoday.intoday.in/story/project-shakti-helped-thousands-of-women-and-also-men/1/195911.html (accessed 5 June 2014).

74 Wang H. *China's National Talent Plan: Key Measures and Objectives*. Brookings. 23 November 2010. www.brookings.edu/research/papers/2010/11/23-china-talent-wang (accessed 17 February 2014).

75 Moavenzadeh J. and Giffi G.A. *The Future of Manufacturing: Opportunities to Drive Economic Growth*. World Economic Forum and Deloitte Touche Tohmatsu Limited. 2012. www.weforum.org/reports/future-manufacturing (accessed 17 February 2014).

76 *Big Demands and High Expectations: The Deloitte Millennial Survey*. Deloitte. January 2014, pp. 2, 3, 8. www2.deloitte.com/content/dam/Deloitte/global/Documents/About-Deloitte/gx-dttl-2014-millennial-survey-report.pdf (accessed 17 February 2014).

77 Sutton Fell S. Top 100 Companies Offering Remote Jobs in 2014. *Huffington Post*. 13 January 2014. www.huffingtonpost.com/sara-sutton-fell/top-100-companies-offerin_b_4578986.html (accessed 17 February 2014).

78 *Millennial Innovation Survey: Summary of Global Findings*. Deloitte. April 2013, p. 9. www2.deloitte.com/content/dam/Deloitte/global/Documents/About-Deloitte/dttl-crs-millennial-innovation-survey-2013.pdf (accessed 17 February 2014).

79 *The Business of Empowering Women*. McKinsey & Company. January 2010, p. 16. http://mckinseyonsociety.com/downloads/reports/Economic-Development/EmpWomen_USA4_Letter.pdf (accessed 17 February 2014).

80 Belsky S. *A Manifesto for Free Radicals: Less Paperwork, Less Waiting, More Action*. 99U. 2013. http://99u.com/articles/7098/a-manifesto-for-free-radicals-less-paperwork-less-waiting-more-action (accessed 17 February 2014).

81 Schawbel, D. How Big Companies Are Becoming Entrepreneurial. TechCrunch. 29 July 2012. http://techcrunch.com/2012/07/29/how-big-companies-are-becoming-entrepreneurial (accessed 17 February 2014).

82 Schawbel D. Why Companies Want you to Become an Intrapreneur. *Forbes*. 9 September 2013. www.forbes.com/sites/danschawbel/2013/09/09/why-companies-want-you-to-become-an-intrapreneur (accessed 17 February 2014).
83 *Report on the Results of Public Consultation on The Entrepreneurship 2020 Action Plan*. European Commission. 29 November 2012. http://ec.europa.eu/enterprise/policies/sme/files/entrepreneurship-2020/final-report-pub-cons-entr2020-ap_en.pdf (accessed 17 February 2014).
84 Kannan S. Bangalore: India's IT hub readies for the digital future. BBC News. 3 September 2013. www.bbc.co.uk/news/technology-23931499 (accessed 17 February 2014).
85 *Startup Ecosystem Report 2012*. Start-up Genome and Telefonica Digital. 3 April 2013, pp. 2, 46–9. http://blog.startupcompass.co/pages/entrepreneurship-ecosystem-report (accessed 17 February 2014).
86 Silver J. East London's 20 hottest tech startups. *The Guardian*. 8 July 2012. www.theguardian.com/uk/2012/jul/08/east-london-20-hottest-tech-companies (accessed 17 February 2014).
87 Moules J. Tech start-ups: Innovation hubs all over world seek to follow Silicon Valley lead. *Financial Times*. 11 June 2013. www.ft.com/cms/s/2/e357a258-b3d2-11e2-b5a5-00144feabdc0.html#axzz375TtyZaq (accessed 17 February 2014).
88 Harris J.G. and Junglas I. *Decoding the Contradictory Culture of Silicon Valley*. Accenture. June 2013. p. 3. www.accenture.com/SiteCollectionDocuments/PDF/Accenture-Decoding-Contradictory-Culture-Silicon-Valley.pdf (accessed 17 February 2014).
89 Lightfoot L. Master's course gives graduates a degree of entrepreneurship. *The Guardian*. 5 March 2013. www.theguardian.com/education/2013/mar/05/master-s-course-give-graduates-a-degree-of-entrepreneurship (accessed 17 February 2014).
90 *The Global Cardboard Challenge*. Imagination Foundation. 2014. www.imagination.is/cardboard_challenge (accessed 17 February 2014).
91 Taleb N.N. *Antifragile: Things that Gain from Disorder.* London: Penguin. 2012.
92 *Preparing for the Silver Tsunami*. Alliance for Aging Research. June 2006. http://archive.is/uCc2n (accessed 1 April 2014).
93 *The Global Burden of Chronic Disease*. WHO. www.who.int/nutrition/topics/2_background/en (accessed 18 February 2014).
94 *World Population Ageing 2009*. Department of Economic and Social Affairs, Population Division, UN. December 2009. www.un.org/esa/population/publications/WPA2009/WPA2009_WorkingPaper.pdf (accessed 18 February 2014).
95 Evans J. *Differences in life expectancy between those aged 20, 50 and 80 – in 2011 and at birth.* Department for Work and Pensions. August 2011, p. 5. www.gov.uk/government/uploads/system/uploads/attachment_data/file/223114/diffs_life_expectancy_20_50_80.pdf (accessed 28 April 2014).

96 Collinson P. Solving Japan's age-old problem. *The Guardian*. 20 March 2010. www.theguardian.com/money/2010/mar/20/japan-ageing-population-technology (accessed 17 February 2014).

97 *Special Report: The World's Oldest Populations*. Euromonitor International. 13 September 2011. http://blog.euromonitor.com/2011/09/special-report-the-worlds-oldest-populations.html (accessed 17 February 2014).

98 Gorman M. and Zaidi A. *Global AgeWatch Index 2013: Insight Report*. HelpAge International. November 2013, p. 17. www.helpage.org/global-agewatch/reports/global-agewatch-index-2013-insight-report-summary-and-methodology (accessed 17 February 2014).

99 *British Public Embrace Retirement Revolution*. Scottish Widows. 10 December 2013. www.business-school.ed.ac.uk/blogs/school-blog/wp-content/uploads/sites/3/2014/01/Retirement-Revolution-Scottish-Widows.pdf (accessed 17 February 2014).

100 *Male over fifties, reliving the swinging sixties*. Mintel. 31 October 2004. www.mintel.com/press-centre/social-and-lifestyle/male-over-fifties-reliving-the-swinging-sixties (accessed 17 February 2014).

101 Smithers R. Happiness linked to financial planning, research shows. *The Guardian*. 16 June 2010. www.theguardian.com/money/2010/jun/16/happiness-financial-planning-aviva (accessed 17 February 2014).

102 Rubin G. Business is booming for the UK's growing army of 'olderpreneurs'. *The Observer*. 5 January 2014. www.theguardian.com/business/2014/jan/05/business-booming-growing-army-olderpreneurs (accessed 17 February 2014).

103 Van de Glinde P. *[Study] The Future Looks Bright for the Dutch Sharing Economy*. Collaborative Consumption. 3 January 2014. www.collaborativeconsumption.com/2014/01/03/study-the-future-looks-bright-for-the-dutch-sharing-economy (accessed 17 February 2014).

104 McGrane S. Car sharing grows with fewer strings attached. *The New York Times*. 26 June 2013. nytimes.com/2013/06/26/business/global/one-way-car-sharing-gains-momentum.html?_r=0 (accessed 17 February 2014).

105 Zeng H. On the move: Car-sharing scales up. *TheCityFix*. 18 December 2013. http://thecityfix.com/blog/on-the-move-car-sharing-scales-up-heshuang-zeng (accessed 17 February 2014).

106 Stone Z. Why the Sharing Economy is Taking Off in Seoul. *Fastcompany*. 17 July 2013. www.fastcoexist.com/1682623/why-the-sharing-economy-is-taking-off-in-seoul (accessed 17 February 2014).

107 Why we share. Peers. www.peers.org (accessed 17 February 2014).

108 Sagar R. 50 shades of green: the collaborative consumption movement. *The Ecologist*. 15 November 2012. www.theecologist.org/green_green_living/1680762/50_shades_of_green_the_collaborative_consumption_movement.html (accessed 17 February 2014).

109 *Goodpurpose Global Study*. Edelman. 2012, pp. 3–4. http://purpose.edelman.com (accessed 17 February 2014).

110 *New Study: Airbnb Generated $632 Million in Economic Activity in New York*. Airbnb. 22 October 2013. www.airbnb.co.uk/press/news/new-study-airbnb-generated-632-million-in-economic-activity-in-new-york (accessed 17 February 2014).

111 James P. Mapping the Future of Education Technology. FastCoexist. 14 August 2012. www.fastcoexist.com/1680348/mapping-the-future-of-education-technology (accessed 17 February 2014).

112 *Skills Supply and Demand in Europe: Medium-term Forecast up to 2020*. European Centre for the Development of Vocational Training. 2010, pp. 10, 13. www.cedefop.europa.eu/en/Files/3052_en.pdf (accessed 5 June 2014).

113 Siegers R. Master in Management's popularity rises in the east. *Financial Times*. 10 November 2013. www.ft.com/cms/s/2/7954e12e-46dd-11e3-9c1b-00144feabdc0.html#axzz2rc14kyZo (accessed 17 February 2014).

114 Kamenetz A. *Exporting Education. Future Tense* (reprint from New America Foundation). November 2013. www.slate.com/articles/technology/future_tense/2013/11/developing_countries_and_moocs_online_education_could_hurt_national_systems.html (accessed 17 February 2014).

115 Singularity Education Group. http://singularityu.org (accessed 16 February 2014).

116 Matthews D. IMF to offer edX Moocs. *Times Higher Education*. 19 June 2013. www.timeshighereducation.co.uk/news/imf-to-offer-edx-moocs/2004924.article (accessed 12 February 2014).

117 *Pensions at a Glance: Retirement-Income Systems in OECD and G20 Countries*. OECD Publishing. OECD. 2011, p. 40. www.oecd-ilibrary.org/docserver/download/8111011ec006.pdf?expires=1398966675&id=id&accname=guest&checksum=1CDF1C9FA72F9D88708A3B5FE3DE2B12 (accessed 12 February 2014).

118 Wilms T. The World in 2033: Big Thinkers and Futurists Share their Thoughts. *Forbes*. 2 August 2013. www.forbes.com/sites/sap/2013/02/08/the-world-in-2033-big-thinkers-and-futurists-share-their-thoughts (accessed 12 February 2014).

119 Haynie D. State Department Hosts MOOC Camp for Online Learners. US News. 20 January 2014. www.usnews.com/education/online-education/articles/2014/01/20/state-department-hosts-mooc-camp-for-online-learners-abroad (accessed 16 February 2014).

120 *The Power of Mobile in Changing Education*. Classroom Aid. 10 March 2013. http://classroom-aid.com/2013/10/03/the-power-of-mobile-in-changing-education-mlearning. Full report by GSMA and McKinsey *Transforming Learning through mEducation*. 5 April 2012. www.gsma.com/connected living/gsma-and-mckinsey-transforming-learning-through-meducation (accessed 16 February 2014).

121 *What is Social Capital*. The World Bank. http://web.worldbank.org/WBSITE/EXTERNAL/TOPICS/EXTSOCIALDEVELOPMENT/EXTTSOCIALCAPITAL/0,,print:Y~isCURL:Y~contentMDK:20185164~menuPK:418217~pagePK:148956~piPK:216618~theSitePK:401015,00.html (accessed 16 February 2014).

122 Economist Intelligence Unit. *The Global Talent Index Report: The Outlook to 2015*. Chicago, IL. Heidrick & Struggles. 2013, p. 5. www.economistinsights.com/sites/default/files/downloads/GTI%20FINAL%20REPORT%205.4.11.pdf (accessed 17 February 2014).

123 Hasle P. et al. *Organizational Social Capital and the Relations with Quality of Work and Health: A New Issue for Research*. National Research Centre for the Working Environment. Denmark. 2007, p. 2. www.arbejdsmiljoforskning.dk/~/media/Praesentationer/PHA-181007.pdf (accessed 17 February 2014).

124 *Social Progress Index 2014*. Social Progress Imperative. 2014. www.socialprogressimperative.org/data/spi (accessed 16 February 2014).

125 Mackey J. and Sisodia R. *Conscious Capitalism: Liberating the Heroic Spirit of Business*. Boston, MA. Harvard Business Review Press. 2013.

126 Happiness should have greater role in development policy – UN Member States. UN News Centre. 19 July 2011. www.un.org/apps/news/story.asp?NewsID=39084#.U-uE216kU8M (accessed 11 August 2014).

127 Williamson M. The serious business of creating a happier world. *The Guardian*. 11 April 2012. www.theguardian.com/sustainable-business/united-nations-happiness-conference-bhutan (accessed 17 February 2014).

128 Kelly A. "Let nature be your teacher": Bhutan takes conservation into the classroom. *The Guardian*. 2 January 2013. www.theguardian.com/global-development/2013/jan/02/nature-teacher-bhutan-conservation-classroom (accessed 15 February 2014).

129 Csikszentmihalyi M. Flow, the secret to happiness. TED Talks (video) 2004. www.ted.com/talks/mihaly_csikszentmihalyi_on_flow.html (accessed 17 February 2014).

130 Confino J. Beyond environment: falling back in love with Mother Earth. *The Guardian*. 20 February 2012. www.theguardian.com/sustainable-business/zen-thich-naht-hanh-buddhidm-business-values (accessed 17 February 2014).

131 Lambert C. The Science of Happiness: Psychology explores humans at their best. *Harvard Magazine*. January–February 2007. http://harvardmagazine.com/2007/01/the-science-of-happiness.html (accessed 6 June 2014).

132 Earth's got talent. *The Economist*. 21 December 2013. www.economist.com/news/leaders/21591872-resilient-ireland-booming-south-sudan-tumultuous-turkey-our-country-year-earths-got (accessed 17 February 2014).

133 Helliwell J.F., Layard R. and Sachs J. (eds). *World Happiness Report 2013*. New York. UN Sustainable Solutions Network. 2013, p. 102 http://unsdsn.

org/wp-content/uploads/2014/02/WorldHappinessReport2013_online.pdf (accessed 17 February 2014).

134 *How's Life? 2013: Measuring Well-being: How's Life? at a Glance*. OECD Publishing. 2013. www.keepeek.com/Digital-Asset-Management/oecd/economics/how-s-life-2013/how-s-life-at-a-glance_how_life-2013-6-en#page1 (accessed 6 June 2014).

135 Diener E., Suh E.M., Lucas R.E. and Smith H.L. Subjective Well-being: Three Decades of Progress. *Psychological Bulletin*.125(2): 276–302. 1999. http://dipeco.economia.unimib.it/persone/stanca/ec/diener_suh_lucas_smith.pdf (accessed 17 February 2014).

136 Jacobs T. Not-So-Tortured Artists: Creativity Breeds Happiness. *Pacific Standard*. 18 February 2014. www.psmag.com/navigation/books-and-culture/forget-tortured-artist-stereotype-creativity-breeds-happiness-74813/ (accessed 13 August 2014).

137 The U-bend of life. *The Economist*. 16 December 2010. www.economist.com/node/17722567 (accessed 17 February 2014).

138 Smedly T. Can Happiness be a Good Business Strategy? *The Guardian*. 20 June 2012. www.theguardian.com/sustainable-business/happy-workforce-business-strategy-wellbeing (accessed 13 August 2014).

139 Achor S. Positive Intelligence. *Harvard Business Review*. January–February 2012. http://hbr.org/2012/01/positive-intelligence/ar/1 (accessed 17 February 2014).

## Chapter 6: Practical trend mapping: focusing on people

1 Schumacher E.F. *Small is Beautiful: A Study of Economics as if People Mattered*. New York. Random House. [1973] 2011.

2 Rubino S., Hazenberg W. and Huiman M. *Meta Products: Meaningful Design for our Connected World*. London. British Interplanetary Society. 2010, p. 71.

3 Sinek S. *Start With Why: How Great Leaders Inspire Everyone to Take Action*. London. Penguin. 2011, p. 41.

4 Haque U. *Betterness: Economics for Humans*. London. Harvard Business Press Books. 2011, p. 29.

5 *Global Risks 2011: Sixth Edition: An Initiative of the Risk Response Network*. Geneva. World Economic Forum. January 2011, p. 6. http://reports.weforum.org/wp-content/blogs.dir/1/mp/uploads/pages/files/global-risks-2011.pdf (accessed 5 January 2014).

6 Kuhn T.S. *The Structure of Scientific Revolutions,* 2nd edn. Chicago. University of Chicago Press. 1972.

7 Schumpeter J.A. *Capitalism, Socialism and Democracy*, 3rd edn. New York. Harper and Brothers. 1950.

8 Schwab K. *The Global Competitiveness Report 2013–2014*. Insight Report. World Economic Forum. http://reports.weforum.org/the-global-competitiveness-report-2013-2014/ (accessed 5 January 2014).
9 *Shared Norms for the New Reality: The Nordic Way.* World Economic Forum Davos. 2011, p. 27. www.globalutmaning.se/wp-content/uploads/2011/01/Davos-The-nordic-way-final.pdf (accessed 13 January 2014).
10 *Corruption Perceptions Index 2013*. Transparency International. http://cpi.transparency.org/cpi2013/results (accessed 9 January 2014).
11 The Nordic countries: the next supermodel. *The Economist*. 2 February 2013. www.economist.com/news/leaders/21571136-politicians-both-right-and-left-could-learn-nordic-countries-next-supermodel (accessed 19 October 2013).
12 Connections Counter: The Internet of Everything in Motion. The Network. Cisco. http://newsroom.cisco.com/feature-content?type=webcontent&articleId=1208342 (accessed 10 February 2014).
13 *Capitalizing on Complexity: Insights from the Global Chief Executive Officer Study.* C-Suite Studies. IBM. 2010. www-935.ibm.com/services/c-suite/series-download.html (accessed 9 January 2014).
14 Goodwille News. Top trends shaping the business world: game changers. 2011. Reproduced with kind permission at http://global-influences.com/social/communication-nation/game-changers (accessed 12 January 2014).
15 Mills C.W. *The Sociological Imagination*. New York. Oxford University Press. 1959.
16 What are scenarios? Shell Global. www.shell.com/global/future-energy/scenarios/what-are-scenarios.html (accessed 15 January 2014).
17 Meister U. (ed.). *Vision 2030: So leben, arbeiten und kommunizieren wir im Jahr 2030*. T-Systems. Offenbach. Gabal. 2012.

## Chapter 7: Practical trend mapping: organizations

1 Edison's Lightbulb. The Franklin Institute. http://learn.fi.edu/learn/sci-tech/edison-lightbulb/edison-lightbulb.php (accessed 12 January 2014).
2 Elliott J. *Secrets for a Good Life*. Bloomington, IN. AuthorHouse. 2011, p. 116.
3 Rubino S., Hazenberg W. and Huisman M. *Meta Products: Meaningful Design for our Connected World*. Amsterdam: BIS. 2010, p. 136.

# Index

Note: figures and tables are indicated in bold

Printed and bound by CPI Group (UK) Ltd, Croydon, CR0 4YY